CHICAGO PUBLIC LIBRARY

S0-BDG-564

HD
7287.96   Housing the poor in
.D44   the developing
H68   world.
1991

$59.95

3/11

R

The Chicago Public Library

Received

BAKER & TAYLOR BOOKS

Housing the poor in the developing world

Housing policies and programmes tend to result from political expediency, rather than a rational and informed analysis of the situation and the demands of individual households for housing. *Housing the Poor in the Developing World* aims to show how methods of analysis can be used to improve efficacy and equity in housing projects and policies, with analysis designed for local circumstances.

This book is aimed at satisfying the need to bring together methods of analysis from several disciplines which can be applied to housing. Each method is presented and illustrated with a case study to show how it can be used to inform housing policy in a wide range of countries in all parts of the developing world.

The methods presented range from intuitive to highly structured and from those dealing with house- and neighbourhood-level issues to those which analyse city- or country-wide issues. Unlike other books in the field, this volume concentrates on the methods of analysis rather than the housing policies and programmes.

The ultimate aim of the book is to illustrate that expediency is not the best way. The book will be of value to students and academics of Geography, Development Studies, Planning, and Urban and Regional Studies. It should also be of interest to agencies and individuals working in the field.

Graham Tipple has spent seven years researching and teaching housing and other urban issues in Africa. He is currently Senior Research Officer at the Centre for Architectural Research and Development Overseas, University of Newcastle.

Kenneth Willis has extensive experience of research involving quantitative techniques in housing and environmental economic fields. He is now Reader in the Economics of the Environment, in the Department of Town and Country Planning at the University of Newcastle.

Housing the poor in the developing world

Methods of analysis, case studies and policy

Edited by

A. Graham Tipple
and
Kenneth G. Willis

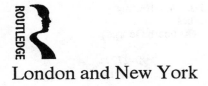

London and New York

To Sue and Pat

First published in 1991
by Routledge
11 New Fetter Lane, London EC4P 4EE

Simultaneously published in the USA and Canada by
Routledge
a division of Routledge, Chapman and Hall Inc.
29 West 35th Street, New York, NY 10001

© 1991 A. Graham Tipple and Kenneth G. Willis

Typeset in Times Roman
by NWL Editorial Services, Langport, Somerset

Printed and bound in Great Britain
by Mackays of Chatham PLC, Chatham, Kent.

All rights reserved. No part of this book may be reprinted or
reproduced or utilized in any form or by any electronic,
mechanical, or other means, now known or hereafter invented,
including photocopying and recording, or in any information
storage or retrieval system, without permission in writing from
the publishers.

British Library Cataloguing in Publication Data
Housing the poor in the developing world: methods of
   analysis, case studies and policy.
   1. Developing countries. Housing
   I. Tipple, A. Graham   II. Willis, K.G. (Kenneth George)
   1944–
   363.596942091724
   ISBN 0–415–05539–3

Library of Congress Cataloging in Publication Data
Housing the poor in the developing world: methods of
   analysis. case studies, and policy/edited by A. Graham
   Tipple & Kenneth G. Willis.
      p. cm.
   Includes bibliographical references and index.
   ISBN 0–415–05539–3
   1. Poor – Housing – Developing countries.
   2. Poor – Housing – Developing countries – Case studies.
   3. Housing policy – Developing countries.   4. Housing
   policy – Developing countries – Case studies.
   I. Tipple, A. Graham.   II. Willis, K.G. (Kenneth George)
   HD7287.96.D44H68   1991
   363.5'96942 – dc20   90–27271
                                                   CIP

Contents

Figures

Tables

Contributors

William Barron is currently a water resources engineer and planner with OXFAM in Cambodia. He is a graduate of the masters programme in water resources engineering at the University of North Carolina at Chapel Hill. He served as a Peace Corps volunteer in Haiti working with the CARE water supply project.

John Briscoe is the Chief of the Water Policy Unit in the Infrastructure and Urban Development Department at the World Bank, Washington DC. From 1981 to 1988, he was on the faculty of the Department of Environmental Sciences and Engineering at the University of North Carolina at Chapel Hill. Many of his publications are concerned with the relationship between water quality and health in developing countries.

Alan Gilbert is Professor of Geography at University College and the Institute of Latin American Studies, London. He has worked for many years on housing and on urban and regional development in Latin America, particularly in Colombia, Mexico and Venezuela. His recent books include *Housing, the State and the Poor: Policy and Practice in Three Latin American Cities* (with Peter Ward, 1985), *The Political Economy of Land* (with Patsy Healey, 1985), and *Landlord and Tenant: Housing the Poor in Urban Mexico* (with Ann Varley, 1990).

Graeme Hardie is Director of Research in the School of Design, North Carolina State University, Raleigh. He qualified and practised as an architect and holds a PhD in anthropology from Boston University. As a researcher in Southern Africa, he has been particularly concerned with the expressive connections between culture and the built environment, particularly among the Tswana people. He is co-director of the newly funded Centre for Accessible Housing and a board member of the Environmental Design Research Association (EDRA).

Gordon Hughes has been Professor of Political Economy at the University of Edinburgh since 1985 and was formerly Lecturer in Economics at the University of Cambridge. His PhD thesis included a cost-benefit analysis of low income housing projects in Nairobi. Since then he has written extensively on applications of shadow pricing and cost-benefit analysis in developing countries and on housing policy in both developed and developing countries. He is currently working on energy, environmental and housing policies in Eastern Europe.

Thomas Klak is Assistant Professor of Geography at the Ohio State University at Columbus. His research on housing policies in the developing world is based on theories of state and aims to help to understand the failure of socially progressive housing policies to assist the poor. His PhD from the University of Wisconsin–Madison examined the workings of the Brazilian National Housing Trust. More recently he has studied the Jamaican and Ecuadorian housing programmes and has been a consultant to the Jamaican Government, USAID, and the Urban Institute, Washington DC.

Jean-François Landeau trained as an economist (PhD from Paris–Pantheon–Sorbonne), co-authored a book on US investments abroad in 1971, and gradually moved into business administration qualifying with a DBA degree from Harvard Business School. Since 1975 his work has focused on corporate finance and business policy. For the last thirteen years he has been with the World Bank, first dealing with industrial development banks and then launching the lending to housing finance institutions. He is currently a financial evaluator of projects across various sectors in the Operations Research Department of the World Bank, Washington DC. He is also writing a book about flaws in the financial markets worldwide.

Stephen Malpezzi is an economist in the Infrastructure and Urban Development Department in the World Bank, Washington DC. Holder of a PhD in Economics from George Washington University, his current research focuses on the relationships between housing markets, urban regulatory environments, and macroeconomic development. He has published a number of papers on housing market behaviour, including *Housing demand in developing countries* (with Stephen Mayo and David Gross, 1985). He is directing a World Bank sponsored research project on Rent Control in Developing Countries.

Xinming Mu received his PhD in City and Regional Planning from the University of North Carolina at Chapel Hill. In 1988, he entered the

Young Professionals Programme at the Asian Development Bank in Manilla, Philippines, where he has held several positions.

Amos Rapoport is Distinguished Professor in the School of Architecture and Urban Planning at the University of Wisconsin–Milwaukee. He has taught in universities in Australia, Great Britain and the United States of America, and held visiting appointments additionally in Israel, Turkey, Argentina, Brazil, Canada, Belgium and India. As a founder of the field of Environment-Behaviour studies, he has focused mainly on the role of cultural variables, cross-cultural studies, and theory development and synthesis. He is the author of *House Form and Culture (1969a)*, *Human Aspects of Urban Form (1977)*, *The Meaning of the Built Environment(1990a)*, *History and Precedent in Environmental Design (1990b)*, and over 200 papers, chapters and articles.

Ann Schlyter is an architect employed as a researcher at the National Swedish Institute for Building Research at Lund. She holds a PhD degree arising out of her series of studies on the long-term development of squatter settlements and improvement areas in Lusaka in association with the University of Zambia. She is also involved in a research project on the housing strategies of women who are heading households in Lusaka and in Harare, Zimbabwe.

Amita Sinha is a visiting Assistant Professor in the Department of Landscape Architecture at the University of Illinois, Urbana–Champaign. She received her PhD in Architecture from the University of California at Berkeley in 1989. She has carried out fieldwork in a village and housing project in Lucknow for the purpose of her dissertation on environmental and social change in Indian society.

Raymond Struyk is Director of the International Activities Centre at the Urban Institute, Washington DC. He holds a PhD in Economics from Washington University in St Louis, and has carried out housing finance, housing need, and housing market analyses in a wide range of developing countries. Among his numerous publications are *Federal Housing Policy at President Reagan's Midterm* (with J. Tuccillo and N. Mayer, 1983), *Finance and Housing Quality in Two Developing Countries: Korea and The Philippines* (with Margery Turner, 1986), and *Assessing Housing Needs and Policy Alternatives in Developing Countries (1988)*.

A. Graham Tipple is Senior Research Officer at the Centre for Architectural Research and Development Overseas (CARDO) in the School of Architecture, University of Newcastle upon Tyne. He worked as a town planner in Kitwe, Zambia, and North Yorkshire, UK before being seconded by Newcastle University to the University of Science and Technology, Kumasi, Ghana, where he taught housing policy and carried out the research on housing policy in Ghana which led to his PhD. He has acted as a consultant to the World Bank and British development companies working in Africa.

Margery A. Turner is a Senior Research Associate at the Urban Institute, Washington DC, and Director of the Housing Research Programme. She holds a Masters degree in Urban and Regional Planning from George Washington University. Her publications include *Housing Market Impacts of Rent Control: the Washington DC Experience (1990)* and *Future US Housing Policy: Meeting the Demographic Challenge* (with Ray Struyk and M. Ueno, 1988), and *Finance and Housing Quality in Two Developing Countries: Korea and The Philippines* (with Ray Struyk, 1986).

Dale Whittington is an Associate Professor in the Departments of City and Regional Planning and Environmental Sciences and Engineering at the University of North Carolina, Chapel Hill. He has also taught at the University of Khartoum, Politecnico di Milano, and the Economics University of Vienna. He has served as a consultant to the World Bank, the Ford Foundation, and other organisations, and is the author of numerous articles on water resources planning and policy in developing countries.

Kenneth G. Willis is Reader in Environmental Economics in the Department of Town and Country Planning, and Co-Director of the Countryside Change Unit in the Department of Agricultural Economics and Food Marketing at the University of Newcastle upon Tyne. He has served as a consultant to British Waterways, the Forestry Commission, Mid-Wales Development, the World Bank, and other organisations; and has researched housing finance and subsidies in Britain for the Joseph Rowntree Memorial Trust. He has a PhD from the University of Newcastle upon Tyne and is the author of several books and numerous articles on environmental and housing economics, and the evaluation of government regional aid.

Preface

For many years, practitioners and students of housing and planning in the developing world have been using methods of analysis developed and written about in contexts other than their own. This book has been put together in the hope that it will help them to relate such methods to housing issues and to discover new methods which they could apply to their local circumstances.

Although each chapter concentrates on the method rather than the case study, no standard format has been imposed so the book cannot be looked upon as an instruction manual. However, it is hoped that enough description of the methods have been included to allow a reader to use them in their own context.

The inclusion of a chapter using a case study in the so-called Republic of Bophuthatswana, should not be taken to indicate any agreement with or acceptance of the apartheid system on the part of any of the contributors, the editors, or the publishers. It is hoped that the case study will assist an equal-opportunity post-apartheid society in South Africa.

The editors wish to thank their colleagues in the Centre for Architectural Research and Development Overseas at the University of Newcastle upon Tyne for their support, and the contributors for their willingness to participate and for their prompt attention to deadlines and modifications. Specific acknowledgements are included in the relevant chapters. Special thanks must go to our wives, Sue Tipple and Pat Willis, for their patience and support, especially as submission day drew near.

A. Graham Tipple
Kenneth G. Willis
Newcastle upon Tyne

1 Introduction to housing analysis and an overview

Kenneth G. Willis and A. Graham Tipple

HOUSING IN THE DEVELOPING WORLD: AN INTRODUCTION

Housing the poor in the developing world is one of the major challenges facing mankind in the last decade of the twentieth century. The challenge is particularly acute in urban areas where populations are projected to grow from a total of less than 300 million in 1950 to almost two billion by the turn of the century; more than 50 million every year throughout the 1990s, an average growth rate of 3.4 per cent per annum. Within the developing world, the growth in urban population is most acute in the poorest countries.[1]

Currently, the major housing problem is the shortage of affordable accommodation for the urban poor; the low-income majority. Over the last three decades, most official housing programmes have failed to reach considerable portions of this group, especially households in the lowest 20 or 40 percentiles of the population. Factors contributing to this failure undoubtedly include the inability of such programmes to provide enough dwellings. As Woodfield (1989) reports, in low-income developing countries during the early 1980s, 61 additional people were born or nine new households were formed for every one new permanent dwelling built. The situation was also serious in middle income countries where population increased at ten times the rate of new permanent dwelling construction. IMF statistics show that governments in developing countries typically spent about 2 per cent of their budgets on housing and community services (Woodfield, 1989: 8)

Formal housing is both scarce and expensive relative to wage levels. Thus, low income households have found niches for themselves in cheaper alternatives often in single rooms in central city rented housing. Rooms in city centre 'vecindades' are the norm for Latin American low-income households (Edwards, 1982). It is estimated that 35 per cent of urban dwellings in Africa are single rooms (United Nations, 1987a), while in some cities the proportion of households living in single rooms rises above 70 per cent (Malpezzi, Tipple and Willis, 1990; Peil and Sada, 1984). Others have built houses in squatter settlements on peripheral or unused land, or found rented rooms in the squatter settlements (Amis, 1987). Some households have even found no housing at all and resort to sleeping under bridges, in culverts, or on central city pavements. Evidence that many squatters and pavement-dwellers have moved out of formal sector housing intended for the poor (Jagannathan and Halder, 1987), only emphasises the problems of combining affordability with the usefulness of houses for low income households in terms not only of the size and quality of the built structures, but also of location in relation to employment and trading opportunities.

Encouraged by the array of national and international initiatives which were stimulated by the International Year of Shelter for the Homeless (1987), the United Nations General Assembly has resolved to formulate a Global Strategy for Shelter to the Year 2000 (Resolution 42/191 of 1988) to encourage governments to assist the current and potential participants in housing production and to create the conditions for adequate shelter production for all demand sectors.

As housing plays such an important part in life, housing analysis has a potentially important role in improving the quality of life at all levels and in every country. Not only can we attempt to understand how the individual house and household are related, at one end of the housing scale, but we can also attempt to analyse the effects of national level policies on housing supply at the other and assess the efficiency and equity of housing policies. The methods used in the analysis of housing have originated in many different fields. As they are scattered throughout the literature on subjects as diverse as economics, sociology, anthropology, statistics, architecture, and management, a newcomer to the field or a busy professional has little chance to review the methods available and assess which might suit his/her particular situation. This volume assembles 12 methods which can be used for issues in the housing sector at a variety of levels. The chapters concentrate on explaining the methods of analysis, using the subject matter of the study primarily as a means of illustrating circumstances in which the method can be used together with its advantages and disadvantages.

ANALYTICAL TECHNIQUES

Information and data must be combined with some technique of analysis to investigate and make sense of any environment, project or policy.

Depending on the context, the data may range from intuitive perceptions on the part of the analyst, to a highly sophisticated set of physical and socio-economic measurements; whilst the technique or method of analysis can vary from some form of intuitive judgement or 'black box' in the analyst's head, to a well-specified and structured quantitative model in which the assumptions and the logical relationships in the model are explicitly documented.

Emulating the physical sciences, social science tries hard to develop a theoretical and methodological perspective to analyse issues; to be as 'objective' as possible in its analysis and in the approach of the investigator; and to quantify the *outputs* of projects and policies as much as possible. Clearly, quantification has progressed further, and is more feasible, in some areas of housing study than in others. Not wishing to exclude any areas of study, whilst at the same time seeking to document a range of methods, the chapters in this book cover a whole gamut of techniques from the intuitive to the highly quantifiable.

The range from intuitive to the highly quantifiable represents both a continuum and a dichotomy, often the subject of fierce debate between seemingly opposing factions. This is brought into focus particularly in a multi-disciplinary subject such as housing where researchers trained in the visual arts (particularly architects) share the field with quantitatively trained economists and statisticians. The researcher who lives among the poor for months listening to their 'bottom-up' perception of housing works alongside the quantitative 'model-maker' who predicts macro-effects of national policies less aware, perhaps, of the effect on the individual.

The issue of subjectivity and objectivity in analysis can be thought of in terms of

1 methods or procedures by which a conclusion has been reached:
   an inquiry is 'objective' when the conclusions do not rely on the personal judgements of the investigator, but rather rely on some formal model which can be quantified, or worked through.
2 the person engaged in carrying out the inquiry:
   an inquiry is 'objective' when the investigator is unprejudiced. An 'objective' investigator is one who has not adopted a particular technique simply because it will produce the desired conclusion he/she wishes to reach; nor manipulated the data; nor excluded part of the data because it does not conform to his/her beliefs.

Table 1.1 documents the four possibilities. A 'subjective' analysis is either one in which personal judgement has been used, or one in which the analyst is prejudiced. Generally, the more 'objective' the

Table 1.1   Objectivity and subjectivity in enquiry and analysis

Method of analysis is	Investigator is	
	Subjective	*Objective*
	(i.e. prejudiced)	*(i.e. not prejudiced)*
Subjective (i.e. personal judgement is used)	1	2
Objective (i.e. no personal judgement is used)	3	4

analyst/inquirer, and the more 'objective' the method of analysis, the better are the results, predictions, and policy recommendations likely to be. Whilst it is generally considered desirable to have investigations of the 'cell four' type in Table 1.1, for some areas or topics in housing analysis this is clearly impossible. An ideal example of this can be found in Chapter 3 by Rapoport and Hardie, in which intuitive judgement is essential at this early stage of developing models to analyse culture change. In such cases 'subjective' methods, in the sense that personal judgements are made, may be the only ones available to tackle the problem. Of course, it is quite possible that prejudiced analysts may employ highly quantitative and 'objective' methods; and that inquiries involving personal judgements may be carried out by unprejudiced and 'objective' inquirers.

The reader may ponder at this stage about what is meant by 'being prejudiced'? How does it differ from looking at a problem from a particular perspective? Self-interest is a key concept in any answer to both of these questions: whether the investigator is interested for him/herself, or is interested from his/her own point of view. Any analyst necessarily looks at things from his/her particular point of view in the sense of bringing to bear his/her particular set of life experiences and cognitive interests. But this interest is different from the investigator's material set of interests: for personal profit, for the benefit of family or friends, or to further some political or religious cause. In reality there is probably a spectrum from zero detachment on the part of the inquirer, to complete detachment.

To view evaluation or analysis as a simple dichotomy between 'subjectivity' and 'objectivity' is perhaps too simple a way of looking at evaluation methods and techniques. A more sophisticated framework, to consider the whole question of analytical method, is provided by

Hammond's (1978) 'cognitive continuum': the notion that most cognitive analysis involves both intuition and analysis in varying degrees (Figure 1.1).

The usual conditions for the application and practice of housing management, and housing as a profession (and most other professions) involve modes four to six. It is here, using these cognitive methods, that decisions are made to build new housing schemes, allocate public housing to certain groups, discourage particular styles of housing, and harass or assist groups who build outside regulations, and many others.

The most intuitive mode, mode six, occurs whenever an individual is operating with minimal support from colleagues or reference to impersonal aids, for example, the housing manager deciding whether to allocate public housing or land to a family; or the single planner or architect designing a housing layout. This mode is very common in the developing world, where professionals are particularly thin on the ground and have few opportunities to consult with others (mode five) because of physical or professional isolation, poor communications, or the exigencies of work. Furthermore, lack of facilities for more structured modes may force intuitive decisions to be made, where, in more ideal circumstances, modes three or four would be favoured.

Intuitive thought involves the rapid and unconscious processing of data. It combines the availability of data from different sources, quantitative and qualitative, by 'averaging' it in some way. Thus the analyst may think about the mid-point of the range, and combine two or more factors by weighting them equally in an additive way. Rarely is it possible for the human brain to combine data in a more sophisticated way, for example, thinking exponentially, or combining several factors in different proportions, to arrive at an answer. Because of this, intuitive thought has low consistency and is only moderately accurate (Tversky and Kahneman, 1974). This has been shown empirically in a number of professions, from medicine (de Dombal, 1984; McGoogan, 1984), to psychology (Kleinmuntz and Szucko, 1984), criminology (Glaser, 1985), and management and investment issues.

This is not to argue that intuitive analysis is irrelevant. Intuitive judgement can be extremely important in analysing issues not amenable to quantitative modelling, and also in 'reflecting' on the results or output of quantitative analysis and methods. Quantitative analysis can, on occasion, throw up very spurious results, so the ability to reflect on such results in context is an art worth cultivating; serious errors can be avoided in this way.

For some (Schon, 1983), expertise resides precisely in the ability to accomplish tremendously difficult synthesising tasks without resort to

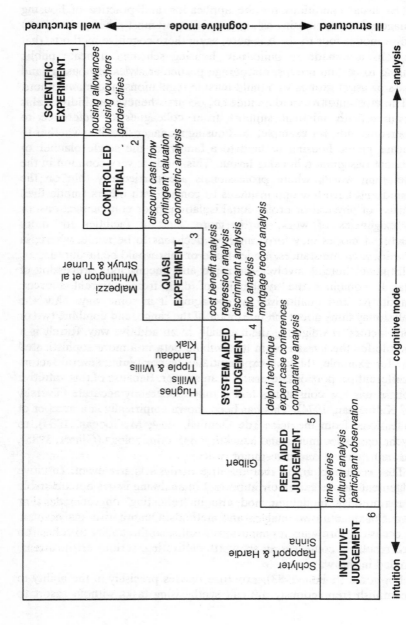

Figure 1.1 Modes of enquiry and analysis: Hammond's cognitive continuum

formal analysis. By increasing an individual's 'knowledge-in-action' and his/her ability to 'reflect-in-action', judgements can be improved. According to Schon (1983: 75–6), when someone reflects-in-action, i.e., reflects on action in action, not after the action is over, he becomes a researcher in a practice context, constructing theory and cases interactively as he frames a problematic situation.

Mode five involves more analysis and a more structured task: a typical example might be a group of planners and/or architects designing a housing estate who meet regularly to review each others' work and discuss particular aspects of the design; or a regular meeting of city housing officers to discuss policy issues; or a case conference of social workers, health workers and housing managers who meet to decide whether a particular family is in need of some specialist form of housing (for example, housing for the disabled).

Peer-aided judgements are more time and resource consuming (for example, case conferences involving many 'experts') but probably allow a fuller and broader analysis, different viewpoints to be explored, and more 'reflection' on the decision. They have been widely used by the United Nations in their 'expert group meetings' to prepare 'state of the art' documents on such topics as the formulation of national settlement policies and strategies (United Nations, 1977), building codes (United Nations, 1980), the development of the indigenous construction sector (United Nations, 1982), the upgrading of inner-city slums (United Nations, 1984), and recently, rental housing (United Nations, 1989 a, b).

System-aided judgements in housing (mode four) are those which use some quantitative methods to analyse projects, policies and cases, and arrive at decisions. The system may consist of data based aids (facts or figures collected in a systematic way and available for analysis in a specific way; for example, to test the association between income level and credit worthiness), ratio analysis, expert systems, discounted cash flow models, or policy capturing equations (regression analysis of house size or condition to income and other variables affecting demand for housing). Such a cognitive mode of analysis is usually at the limits of professional practice and everyday application in the real world.

The usual conditions for the acquisition of knowledge, as distinct from its application, involve modes one to three. It is in these modes that theoretical and methodological research is advanced. Analysis here, at the other end of the continuum from professional practice, also has the opposite features: analytical thought is slow, conscious and consistent, and is usually quite accurate.

Scope for quasi-experiments, but more especially controlled trials and scientific experiments, are much more limited in the social sciences than in the physical (physics and chemistry) and biological sciences. It is much more difficult in the social sciences to alter one variable to observe its impact, whilst holding all other variables constant. Nevertheless, many researchers have tried hard to conduct their analyses in modes one to three, to do just that.

Whilst laboratory experiments (mode one) have been conducted in some areas of economics (Smith, 1980), none to date seem to have been associated with the economic issues associated with housing. Scientific experiment as a method is probably severely limited in its potential application to housing. However, controlled trials and quasi-experiment type methods of analysis (mode two) are much more common.

The 'model villages' of Great Britain; New Earswick, York (Rowntree); Saltaire, Bingley (Salt); Bourneville, Birmingham (Cadbury); and Port Sunlight, Liverpool (Lever) in nineteeth century England, can be thought of as early controlled housing trials. Houses were built to certain standards with planned streets; public buildings provided; and establishments selling liquor often excluded, to improve the 'moral' well being of inhabitants. Later examples from the late nineteenth and early twentieth century included the Garden City concept of Ebenezer Howard, with its fruition in terms of Letchworth and Welwyn Garden City. This tradition permeated through the twentieth century with the New Towns Act 1946 in Britain. A large number of experiments in house design have taken place, especially in the Netherlands; a country that has earned the reputation of the housing laboratory of Europe. Experimental housing in Adelaide Road, London, by Wilkinson and Hamdi of the GLC (Hamdi, 1978) and Papendrecht, Rotterdam, by van der Verf of KOKON, allowed tenants to make their own floor plans within the dwellings (van Rooij, 1978, and see back numbers of *Open House International* for many small-scale experiments).

In the developing world, the modern movement in architecture, with schemes such as Chandigarh (India) by Le Corbusier (Sarin, 1982) and Brasilia by Lucio Costa attempted to improve the quality of people's lives by means of improved design. On a smaller scale, Fathy's (1973) experimental village of New Gournaa used traditional Nubian mud vaults to demonstrate that buildings could be low cost, spacious, cool, and beautiful all at the same time.

Examples of controlled trials (mode two) from the United States are the New Jersey income maintenance experiment, the negative income tax and labour force experiments, and the cash housing allowance experiments (Feber and Hirsch, 1982; Carlson and Heinburg, 1978) in selected areas.

There are a number of problems associated with social experiments or controlled trials in housing. First, such experiments are often extremely costly, running into millions if not hundreds of millions of pounds. As such it is a method of research not likely to be open to many researchers, except those employed by governments. Second, there is the time element: people do not adjust simultaneously to changes in the economic environment; such adjustments require time. So impacts of experiments may not be observable until some time in the future, perhaps several years. Third, the fact that people know they are being experimented upon can have an effect in itself, independent of the nature of the economic, policy or variable which has been altered. People's behaviour and responses can alter simply because they know they are being observed, as de Dombal *et al.* (1974) note.

Quasi-experiments (mode three) in the social sciences are conducted by finding some counterfactual position against which to compare the situation under investigation; that is, to compare the situation *with* the policy or project to that which would occur *without* the policy or project. In this way the *net* impact of the policy can be determined. Typically in economics, such a counterfactual position is determined by some direct method, such as contingent valuation; or an indirect method, for example, using a hedonic price technique to estimate what a house subject to rent control would rent for in an uncontrolled market.

Contingent valuation methods are based upon asking people directly about their preferences and values, as in willingness-to-pay surveys, and in bidding games. The respondent is asked to imagine some hypothetical counterfactual position and then asked to estimate how their behaviour would change if such a situation occurred. For example, if a piped potable water was supplied to your house, how much per week would you be willing to pay? All contingent valuation surveys involve questions about hypothetical situations. The main weakness of any contingent valuation analysis, or other survey method based upon asking people about their preferences, lies in the variety of biases to which the technique is prone. Careful framing of questions may overcome many of the problems with this technique (Whittington, Lauria and Mu, 1990).

Indirect methods, of which hedonic price methods are examples, avoid bias in individuals' responses to direct and hypothetical questions by using observed or actual expenditures for goods and services in a related or alternative market. With the hedonic price technique, a housing unit is viewed as a bundle comprising many characteristics; for example, number of rooms; toilet, bathroom, kitchen facilities; garden area; location relative to urban services in the community; and so on. Each characteristic is assumed to have its own price which can be estimated with regression analysis; the hedonic price index is an attempt to consider the marginal effects of the different characteristics of the heterogeneous good on the price of that good.

Hedonic price methods underpin many housing research projects both in developing and developed countries. They form the basis of Chapter 10 by Struyk and Turner, while cost-benefit analysis of housing policies or projects often rely on hedonic prices as Hughes points out in Chapter 13. Controversy associated with the hedonic price technique centres on biases in the regression coefficients, and thus the extent to which the technique can produce a 'true' value of each characteristic of the good. Multicollinearity (see Chapter 9) in hedonic models often makes it difficult to estimate reliably coefficients of individual *characteristics*; but this does not necessarily mean that overall predictions of rent or price for particular *units* are unreliable or biased.

In this sense methods (indirect or otherwise) which are based upon people's actual behaviour in the market place may be, but not necessarily, more robust. Thus, what a housing unit would rent for in the absence of rent controls can be inferred from what units rent for in the uncontrolled sector, correcting of course for differences in attributes between the two properties. This may provide a better method of estimating the subsidy from rent control, than asking a household, by a contingent valuation question, how much it would be willing to pay for the current unit if rent control did not exist. Long exposure to rent control may change tenant's perceptions of the uncontrolled market rent of the house, thus biasing the response to a contingent valuation question. However, the use of housing indices are not without problems, associated with the composition of the housing stock (Fleming and Nellis, 1987), or econometric specification issues, such as omitted variable bias, the functional form of the model, etc. (Graves, *et al.* 1988).

All techniques have their strengths and weaknesses. It is the researcher's responsibility to assess these, and to choose the most suitable method for the analysis to be undertaken, bearing in mind resource constraints. The purpose of adopting a technique is to improve understanding of the project or policy in question. But much depends upon the investigator who is using the technique: on his or her ingenuity. The aim of this book is merely to provide a flavour, to whet the appetite for further research, investigation, and development of techniques and their empirical application. However, it should be remembered that even the most careful application of a technique will not prevent all errors of judgements nor all mistakes. The application of system based aids in Landeau's chapter on ratio analysis makes this point. But the rigorous application of techniques, by improving discriminatory power, will minimise errors of judgement in projects and policies, and improve decision making.

Attached to the 'cognitive continuum', by way of illustration, is what the editors perceive to be examples of the different types of analysis deployed and documented in this book. A few other examples have also been included to fill out the picture. But whilst trying to be 'objective', *we* recognise that the method of analysis employed to assign chapters is firmly rooted in mode six: intuitive judgement! Hence high consistency cannot be claimed, nor any great accuracy: concensus among chapter authors, not to mention readers, with the assignment in Figure 1.1 may be low. In addition, while parts of each analysis in any chapter may equate with the assignment in Figure 1.1, other parts may be better placed in quite another mode. Thus discounted cash flow analysis can be used in a system aided judgement; but as a technique, it can also be used to carry out quasi-experiments, by changing some of the assumptions incorporated in the model. Nevertheless, the figure as it stands should help the reader in structuring the book, as well as understanding the arguments in this overview.

The methods of analysis featured in this collection range widely in the point from which they view housing; from the fine variations of environment and the intimate relationship which occupants have with their housing environment (for example, in Chapter 3) to the concern with suitable housing policy to cater for typical or median households within a range of economic and social conditions (in most of the later chapters).

Chapters 2 and 4 are variants of qualitative research in which the extended interview is the basis for data gathering and provides a rich mixture of hard data (for example, on household size) and pertinent anecdotes and impressions. The richness of insight which can be gained from such techniques, which are rooted in urban anthropology, are at the heart of their appeal for some analytical tasks, and as a complement to the more quantitative methods which can easily miss vital attitudinal issues (Ward, 1982). Contributions to understanding housing in the developing world using participant observation techniques include the controversial work of Oscar Lewis (1959, 1968), autobiographical accounts of squatting (de Jesus, 1962), and Wikan's (1980) captivating study of the life of poor women in Cairo. Sinha's participant observation (Chapter 2) and Schlyter's longitudinal study (Chapter 4) both examine the techniques in the context of the quality of life for the urban poor in specific, fairly small areas, in order to assess the impact of past and current housing policies. Schlyter's longitudinal study on squatters in Lusaka considers *inter alia* the logistics of comparative research through time.

The inter-relationship between design and culture, and the recent growth in attention directed to the decline of traditional cultures in the developing world, has generated a rich academic debate typified by the work of Rapoport (particularly 1969a; 1977; 1990a), Oliver (1969; 1971; 1975; 1987), and others. The part of this work which most concerns us in this volume is the necessity for planners and architects responsible for housing designs and residential layouts to be aware of the cultural requirements of their clients. This has relevance not only to what happens on the plot but also to the larger spatial plans suitable for culture-specific house and neighbourhood types. For example, where culture demands multi-household compound houses, policies on urban layouts should be different from those required for single household bungalows or multi-storey apartments. Similarly, the resulting tenure patterns and residential densities will be different.

The pioneering qualitative and intuitive research of Rapoport and Hardie (Chapter 3) shows how a researcher can determine which aspects of a culture are essential to its continuity and how, by building with this in mind, housing can be supportive to its occupants in times of rapid cultural change. Such work is highly culture specific, each culture will need examining individually, but this chapter lays the ground rules for understanding environment-behaviour relations in a way which will allow culturally supportive housing policies to be formulated.

Gilbert's Chapter (5) recalls an earlier tradition of analysis; comparing characteristics across nations, cultures, or periods to learn more about each from such comparisons. In the urban field there has been a growing stream of comparative work since the 1960s with Breese (1969) and others leading where Friedmann and Wulff (1976), Walton and Masotti (1976), and others have followed. The literature on housing in the developing world has also benefited from a stream of comparative research, including such notable contributions as Peil (1976) on squatter settlements in Africa and Schwerdtfeger (1982) on housing in Islamic cities.

Gilbert shows how comparative analysis, comprising in-depth case studies of two or more cases, can provide a rich vein of insight into the processes which underlie housing conditions, forcing the researcher to recognise diversity in both cause and effect. In the light of his twelve years' experience of comparative research in Latin America, he stresses the importance of balancing the requirements of the comparative study with the concerns and strengths of the local teams.

The chapters by Klak (6) and Landeau (7) point up the richness of insight available from manipulating existing data sets. Subject to the difficulties attached to gaining access to government or financial institution data sets, for which Klak's advice is invaluable, much can be learned from them which can inform housing policy. For example, they contain 'smoking gun' evidence on the equity and efficiency of allocation procedures and how closely performance approximates to policy in housing finance provision. In addition, through Landeau's ratio analysis, the relationships between what a household borrows, its income, the debt service payments, the house cost, and the initial deposit paid throw valuable light on the nature of affordability and what a household is willing to pay for its housing.

The use of discriminant analysis to trace the differences between households with different housing characteristics is demonstrated by Willis and Tipple (Chapter 8) in relation to tenure and the use of services. By constructing a model using variables suggested by tenure analysis elsewhere, and the use of readily available computer software, the causes of differences in housing consumption can be shown and the likely effects of changes in circumstances can be demonstrated.

Multivariate analysis frequently centres on estimating linear relationships between variables through regression analysis. While discriminant analysis is one form of linear model, some of the many forms which linear and regression models can take are described in Chapter 9. The chapter points to the powerful nature of the regression technique as a means of data analysis, compared with the simple cross-tabulation, in identifying which variables 'explain' variations in the item or issue under consideration. However, the chapter emphasises within the scientific tradition of logical deduction, the necessity of carefully ensuring that all the assumptions must be controlled or corrected if valid conclusions are to be drawn. In addition, it outlines techniques of analysing qualitative variables. These are capable of greater extension in housing analysis. Ordinary regression is illustrated by way of an examination of the determinants of overcrowding; while house condition provides an example of how qualitative data can be modelled by regression analysis.

The recent growth in the use of hedonic models to assess the economic effects of subsidies and other government interventions is reflected in Struyk and Turner's description of how the impact of rent controls can be measured (Chapter 10). The technique allows a researcher to estimate the market value of a good by means other than its price (which is distorted), and from this to calculate the costs and benefits of the price distortion to the various parties involved. Such analyses show that benefits may not be as progressive as policy intended; indeed, in the case described (and elsewhere, see Malpezzi, Tipple and Willis, 1990), benefits are shown to be regressive; being of greater benefit to the higher income groups.

While Landeau provides insights into finding a client group's demonstrated willingness to pay for housing, the amount which people would be willing to pay for a service which does not yet exist is more problematic. Whittington, Briscoe, Mu, and Barron (Chapter 11) show how direct questioning may not be as unreliable as some authors have previously suggested. Their case study in Haiti is directly related to water supply which is one of the bundle of goods and services which, put together, constitute housing.

Malpezzi (Chapter 12) shows how the use of discounted cash flow analysis can compare costs and benefits of different housing programmes for owners or renters, and allow the comparison of different components of the programmes. It replicates, in a highly controlled and structured manner, the 'back of the envelope' calculation made by a developer when deciding whether to go ahead with a scheme. By setting up such a model on a micro-computer, keying in locally relevant values, and changing them to reflect different policy decisions, a professional can demonstrate to him/herself, and indeed to a local decision-maker, the monetary effects of housing programmes (for all the parties involved) and their sensitivity to particular components and policies.

Cost-benefit analysis (CBA) is often regarded as an all-embracing technique, derived as it is from welfare economics. It appraises a decision from the point of view of society as a whole and it incorporates the external effects of a project or policy into the (financial or discounted cash flow) analysis. The technique is also capable of evaluating a project or policy from a general equilibrium perspective, as well as in a partial equilibrium framework. CBA usually requires the re-estimation of market values in terms of an opportunity-cost perspective or of benefits foregone elsewhere. Hughes (Chapter 13) outlines the principles by which this should be undertaken for housing projects in the developing world, and provides a flavour of the many nuances involved. He is also careful to distinguish between efficiency and distributional benefits which may often be overlooked in appraisals. CBA is not regarded as a technique to be applied to a housing project which has been rejected in financial terms just to see if a decision to go ahead can be justified on social terms; nor should the technique be restricted to evaluating a project after it has been completed. Rather, Hughes argues, CBA should be integrated into the design and planning stage of a project to improve the design and balance of components within it, before it is completed, to ensure the best use of resources and the maximisation of benefits.

The latter concept applies to most of the methods of analysis outlined in this volume and takes us back, full circle, to the chapters by Sinha, Rapoport and Hardie, and Schlyter, which emphasise the importance of design and the need to assess preferences of tenants and occupiers in creating inhabitable housing. Qualitative and quantitative approaches to a project or policy can be complementary and can be mutually applied, with the analysis under each technique drawing upon the application of other techniques. This is more likely to ensure not only good design and management of space, but also the maximisation of the project or policy's financial and economic viability, and its overall success.

While the case studies are felt to be of secondary importance in this volume (the techniques being the centre of our attention), efforts have been made to provide examples from many parts of the world and in countries with widely differing economic contexts. Latin America is well represented with cases from Colombia, Mexico, Chile, Colombia, and Venezuela in Gilbert's comparative analysis, Jamaica and Brazil for Klak's mortgage records, and Haiti in Whittington *et al.* Sub-Saharan African case studies are featured by Rapoport and Hardie (in the so-called independent homeland of Bophuthatswana), Schlyter (Zambia), Malpezzi, Willis, and Willis and Tipple (all using Tipple's data sets on Kumasi, Ghana), and Hughes (Kenya and Tanzania). North Africa and the Middle East are represented by Landeau's chapter on Tunisia, and Struyk and Turner's case study drawn from Jordan. Asian case studies are taken from India (Sinha) and Malaysia (Malpezzi).

The book aims to be catholic (but not exhaustive) in its coverage of methods of analysis, with something for the numerate and the non-numerate reader. Just as many disciplines contribute to the understanding of housing in the developing world, so each technique has a place in the spectrum of housing analysis. It is the editors' hope that at least one of the methods will prove invaluable for each reader and will further the cause of increasing efficiency and equity in the allocation of scarce resources to housing the poor in the developing world.

NOTE

1 Countries are defined by (United Nations, 1987: 50, 60) and by Woodfield (1989) as follows:
  low-income; those whose per capita GNP, averaged over 1980 to 1983, was under US$400, in 1975 dollars;
  middle-income those with between $400 and $1400 per capita GNP.

2   Participant observation

A study of state-aided self-help housing in Lucknow, India

Amita Sinha

INTRODUCTION

Participant observation techniques illustrate how qualitative research methods can be used in studying the built environment, more specifically low income housing in developing countries. Reasons for undertaking qualitative research in this field are various and, while there are shortcomings to the method, modifications of the anthropological research methods can be made to suit the purposes of environmental design research. Fieldwork carried out in Northern India is used to illustrate the technique. Finally, the advantages and disadvantages of studying the observer's own society and its physical environment are discussed.

Each discipline must find its own way of doing research, formulating and advancing theory. The model followed by environmental design research, especially in evaluation of the built environment from the user's point of view, is a causal deterministic one, derived from psychology. This paradigm has been shaped by logical positivism which emphasises testing and quantification rather than interpretation. In a quest for the ability to generalise, the testing and quantification may be of concepts which have little validity. Very often the lack of meaning can arise from an inability to interpret the data in terms of a social and political structure. A superficial examination of user behaviour can lead to neglect of social and cultural norms which guide behaviour. The emphasis on causality in terms of independent and dependent variables is at the cost of understanding the mutual influence of a number of factors.

If architects are to be involved in the research and evaluation of the built environment, this prevailing model of research is unsatisfactory. By their very training, designers are more inclined to do interpretive work. A research outlook is needed which can adequately describe the qualities of the fluid interaction between people and the physical environment, and interpret the built environment as a symbol of the cultural values and social order of its inhabitants. This will entail moving away from a deterministic model, in which the ability to generalise and statistical corroboration is emphasised, to one which seeks to understand the subjective experience of individuals and views unique events as keys to useful insights.

QUALITATIVE RESEARCH IN HOUSING EVALUATION

What form should this research take? Evaluation of the congruence of housing to its occupants' way of life and of resident satisfaction, are applied social research; and are often done to inform policy and design guidelines. They may start off as an attempt to 'fix' problems, investigate unmet needs, or lay down broad guidelines for further environmental change but they are a form of action research, done in the field, in the context of the environmental setting investigated. However, the test of their efficacy lies in the way they are carried out. Standardised survey instruments, which permit easy replication of the research process, often result in little other than a listing by the residents of what is wanting in the environment. They fail to capture the meaning and significance of housing for its inhabitants or the larger political and economic structure which lies behind that form of shelter.

Fieldwork done in social and cultural anthropological traditions can be suitably modified for studying the built environment as a manifestation of values, ideas, and attitudes of the society. It can be used to study the associations that people bring in perceiving and using new forms of housing and the transformation in their life-styles that this brings about. Qualitative research methods can be used to study how people become conscious of, give meaning to, and relate to, the built environment. Participant observation, used in anthropology and other fields such as education, is particularly suited to the investigation of such issues. It permits a flexible research design, enables considerable exploration in the field before settling down to a particular area of research, and allows for unexpected phenomena which may prove to be significant. There is considerable scope for the generation of working hypotheses in the course of fieldwork. Thus grounded research, in the sense that field data itself shapes the enquiry, can be carried out. Research formulation, data collection, and analysis, occur in a cyclical fashion instead of following a linear path.

Evaluation of housing often arises out of the necessity to make space more useful. This form of context-embedded enquiry is congruent with the nature of participant observation. It seeks to understand a particular setting or a substantive problem rather than proving a theory with abstractly defined propositions. It is particularly useful in a field without a well-developed theoretical base. The phenomena which are studied in evaluation of the built environment, namely the interaction between people and spaces, are constantly changing, making the theoretical framework weak in prediction. In this form of research, the researcher does not embark upon fieldwork with a tight research design but can choose direction as his/her understanding of the situation grows and new opportunities in the field present themselves. The researcher should be armed with an orientation that will enable him/her to focus on phenomena that deserve more attention than others.

There is great potential for applied social research using a qualitative open-ended research strategy which can be geared towards answering policy questions or making future design decisions. In housing, partici-pant observation is useful in the evaluation of recently implemented schemes and existing low income or squatter communities for the purposes of upgrading; or for studying resident needs, their access to resources, and the physical solutions. It is particularly useful in investigating the composition of community organisations in existing settlements which can be harnessed to assume the duties of local housing authorities, arranging assistance for individual families, and even resolving the problems involved in site layouts and house construction.

The shortcomings of using a qualitative research method such as participant observation include the lengthy research period. An initial exploration or a pilot visit has to be followed by months and even years spent in the field. This form of research is also more heavily influenced than other research methods by the researcher's bias, world-view, and personality. There is also a greater likelihood of its outcome being perceived as a personal interpretation of events.

A combination of methods may be used: interviewing, observations of behaviour, plotting life histories with accompanying changes in the environment, recording of physical data such as house and neighbourhood layouts, building materials, furniture arrangements, etc. These, of course, have to be guided by an orienting theory derived from a conceptual framework in which the physical environment plays a salient role. Certain pitfalls should be avoided; for example, standardised research instruments which seek to eliminate researcher's bias, and endless recording of behavioural usage of space for statistical verification. Pitfalls may occur if certain places in a housing project are used only on ritual occasions and may be visited only a few times during the year. That does not detract from their unique significance and meaningfulness for the residents. The elimination of researcher's bias in recording data also eliminates the skill and empathy that a researcher brings in recording and interpreting a situation. Surveys and formal interviews may not yield consistent and true answers on the part of the respondents. As William Whyte (1984: 69) points out:

> If we insist on asking people what is going on and why they are acting as they do, at best we get formalised explanations; the interpretations people give to outsiders. If we are successful in establishing a social base, we may find quite suddenly that we have broken through the superficial level, that we begin to see patterns and movements which were not evident before, and at last we begin to get a vision of what it is we are really studying.

Hence it is necessary to build up a rapport with the informants and interview them on as many occasions as time will allow. The physical environment, especially housing, forms the context for everyday behaviour and may be taken for granted to such an extent that the user is hardly aware of his/her surroundings. It may become part of the person's subconscious realm surfacing only when it poses a risk or becomes difficult to use. Therefore, residents may be inarticulate when asked about the built environment. Thus informal interviewing and unstructured observations should be used depending upon the circumstances. This combination or triangulation of research methods will yield more complete and richer information than can be obtained by using each method singly. A similar reasoning can be followed for correcting informant bias by comparing one informant's account with that of another.

CASE STUDY: HOUSING IN LUCKNOW

In Lucknow, as in other Indian cities, the government seems to be concentrating its resources on large-scale housing projects on the fringes rather than on development of new towns or the upgrading of the existing housing stock. The housing projects are thus helping to disperse people and activities from the core of the congested city. The government is, in a sense, pre-empting the squatter colony settlements which have a tendency to spring up on the outskirts of a city. However, the beneficiaries of the government intervention are rarely families belonging to the same socio-economic group as the squatters. The reasons for development in projects on the urban periphery include easier acquisition of a large amount of agricultural land. Most of the residential development in Lucknow has taken place in the form of a corridor along axes of transportation to other cities.

In the Indiranagar housing project, a decentralised local institution, the local Housing Board, undertook land acquisition, the designing and building activities, and the allocation of the housing units. HUDCO (Housing and Urban Development Corporation), which is the main national level financial institution in India, financed the planning and building activities. The framework for policy planning and implementation are laid down at the national level.

The location of the housing project on the city periphery removes the low income group from sources of employment and close family and social ties within their neighbourhood. The high levels of centralisation and bureaucracy characteristic of state provided housing induces inefficiency, high administrative costs, and opportunities for corruption in the processes of allocation and approval of alterations and additions. This study concentrates on the part of this project which is targeted at the economically weaker section (EWS), the poorest sector of the Indian urban population. It is constructed and allocated with no profit margin and benefits from a subsidy from profits on high and medium income housing and sale of plots meant for commercial purposes.

As Burgess (1985) points out, state self-help housing solutions cannot effectively duplicate the economies of the squatter settlement. Though there is a certain economy of scale derived from mass production, what is achieved in self-help squatter communities, i.e. employment to small-scale building contractors and saving of financial resources through the contribution of family labour, does not occur in state-aided self-help housing projects. Given the other pressing demands on government expenditure, such housing is likely to remain outside the effective demand of the vast majority of the population. The housing in this case study is, as we will see, no exception to this.

The institutional structure of initial down payment, regular monthly payments, and the penalties attached to default, does not fit in with the irregular income of construction labourers, street vendors, and others in typical marginal urban occupations. The time gap between registration for a house and its actual allocation is about five to eight years and, during this period, the economic condition of the household may change for better or worse. The EWS housing in most cases becomes appropriated by the low income group which is the established working class with regular income.[1] In addition, there are absentee landlords who have obtained more than one house with the sole purpose of renting out units at high prices; about 10 per cent of the houses surveyed were rented units.

As soon as the state develops a piece of land, its price rises steeply, putting it out of the reach of the urban poor. They are thus denied the land to which they would have access in a squatter settlement. Speculation is encouraged by high state subsidies which cause the house to be priced much lower than in the open market. The ownership of land means that the family is free to sell the house to wealthier households at a profit. However, the result is that state subsidised housing does not stay with the economic group for which it was meant. On the other hand, if the EWS household retains the plot, their tenure is insufficient to obtain access to financial and mortgaging institutions. While the planning procedure shifts the burden of harnessing resources and building activity on to the residents, the constraints built into the incremental building process (including a plan to be rigidly followed) lead them to break regulations and bribe their way through the bureaucracy.

Though no official figures are available, in the author's survey of 50 households on EWS plots in Indiranagar it was evident that the income of 90 per cent of the residents far exceeded the allowable income range in EWS housing.

The Indiranagar layout was planned using an efficiency model in which the main criteria seems to be to produce a maximum number of housing units with the least expenditure of resources (see, for example, HUDCO, 1982). The site (which has villages dotting the agricultural landscape) is treated as a blank slate (Figure 2.1). Though the villages and hamlets were allowed to remain, no move was made to incorporate them in the project, reducing them to isolated slum pockets in the newly constructed housing estate. The layout is dominated by row housing clustered around open unpaved squares. Cross-subsidisation is achieved through a mixture of income groups. The housing colony is divided into areas for the high income group, the middle income group, the low income group, and the economically weaker section (EWS). The household distribution in Indiranagar is as follows: high income, 5.08 per cent; middle income, 16.42 per cent; low income, 27.09 per cent; EWS, 42.10 per cent; and land plots, 9.31 per cent.

The EWS houses in the study are provided in the form of a core on each plot consisting of one room, kitchen, WC, and open space for addition to the house. It can be called self-help housing only to the extent that the decision to undertake additional construction rests with the family.

The designer produces a number of standardised housing models dominated by the concept of efficiency, minimum costs and rapid installation.[2] Frequently the designer has no real experience of staying in a large family with constricted space. Usually he or she comes from

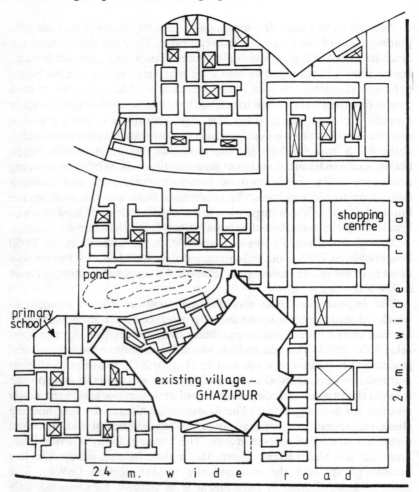

Figure 2.1 Plan of a housing colony in Indiranagar, Lucknow

an urban, middle-class background with no understanding of traditional dwellings and ways of life they foster. There is no recognition of the critical economic functions that the house fulfils through accommodating the extended family, or allowing subletting, provision of open area for market gardening, poultry keeping, and space for a shop. Shortcomings in the design are excused as being necessitated by financial constraints. User needs receive the lowest priority from the housing agency whose main goals seem to be cost recovery, allocation procedures, and approval of alterations in the dwellings. Thus, the housing agency searches for uniformity, standardised products, and ease of administration; local variations in physical and social conditions are ignored. No programming or post-occupancy evaluation is done nor is it seen as essential.

There is a belief that higher standards than those provided in EWS housing will result in a higher socio-economic group taking possession of the house.[3] When housing is state provided with different facilities according to the targeted income group, questions about fair distribution and provision of adequate standards are raised. Should a family of six be forced to live in a one-room house, in a neighbourhood which has relatively good amenities? The differences result, of course, from the families' differing ability to pay, but it can be argued, even in low income housing, there is a certain threshold below which the quality of shelter should not be allowed to drop. A 3 metre by 4 metre room, as in EWS housing, serves the definition of only minimal shelter, with no provisions for privacy and varied activities that make up the structure of everyday life in a household. As Burgess (1982) points out, Turner's (1968) emphasis on minimum standards will reinforce the *de facto* duality of standards between what is permissible for middle and upper income groups, and what is permissible for low income groups, and sanctify it by law. A distinction should be made between reduced standards in space and in services. Service standards can be improved gradually while space cannot be increased in the tight housing layout (consisting of back-to-back row housing, with the provision for adding an extra room in front and building on the first floor only). Low space standards cause the family to cover 100 per cent of the site when additions to the house are made.

Participant observation was used in the fieldwork to study the impact of the EWS housing project on residents in Lucknow. This additionally involved investigating traditional forms of housing and the lives lived in them to allow interpretation of environmental change at the micro-level of the house and neighbourhood in terms of its social implications. A fairly open-ended research strategy was used, relying on the co-operation of a number of households in the new Indiranagar housing project and a village on the urban fringes of the city of Lucknow. The method of analysis adopted involved constant comparison in every relevant aspect between the rural and urban communities, representing two points in a continuum of change. During the course of research, pragmatism linked with situational constraints determined the choice of methods.

The researcher using participant observation was faced with a decision on the extent to which he or she should participate in the two communities. Obviously not everything in the structure of everyday life can be recorded or queried. Data gathering has to be selective, in this case towards those aspects of social life in which the physical environment has a tangible role to play.

The main research technique in this case was in-depth interviewing. The questions were directed towards how space is used. What is an 'ideal' environment? How did a dwelling or part of it come to be built the way it exists? A standardised survey form was not used, partly because many of the research issues became clear only as fieldwork progressed and partly because it was felt that the many significant nuances characterising people's shift in attitudes and values as a result of changing environments would not be captured by surveying. Therefore, in-depth interviews (guided by an interview schedule and lasting over a couple of hours during one or more visits) were felt to be potentially more rewarding. The verbal data were supplemented with sketches of house plans and furniture layout and informal behaviour observations made during the time the researcher spent in the setting.

Contact with the householders was made through residents of the locality selected for study who were willing to introduce the researcher to their neighbours and acquaintances. The only criterion used for the selection of informants was that they should be inhabitants of the site and be willing to be interviewed. The first site (village) was selected after a number of false starts (residents in a couple of other villages refused to be interviewed after an initial meeting) while the second site was part of the housing project in which the author was resident throughout the period of study. The sample was clustered in the low income zone in both the settings. However, the sampling was purposive rather than random, since the interviews took place as a result of personal introduction.

The research methodology in this study was not geared towards discovering causal relations between environmental and social change but towards interpreting environmental change in its social dimensions. The theoretical perspective or framework, namely, the study of dialectics between socio-cultural issues and the built environment at the residential scale, guided the framework. This perspective helped to draw the conceptual boundaries of research.

The myth of a detached researcher cannot always be maintained in real-life settings. The informants respond (or do not respond) to questions within their own cultural framework. Many questions such as 'What is your "ideal" house like?' or 'How do you use the spaces inside the house?' may be treated as if the answer was self-evident; especially if the researcher is a member of the same culture and, therefore, assumed to understand their unarticulated aspirations and assumptions; that is, not being enough of an outsider to deserve an explanation of the everyday course of life. This can be one of the disadvantages in studying one's own society.

In this case, there appeared to be a tendency on the part of the informants to tailor their answers in accordance with their perception of what the interviewer expected to hear. The villagers were anxious to project a favourable image of the village. In many cases, the responses were not intended to communicate reality but to impress. However, this became clear only later with increasing familiarity with the village and its residents. In such a situation, a structured questionnaire survey would probably have yielded a very distorted picture.

On many occasions the writer felt that she was intruding into the privacy of informants' lives by her questions regarding the household and family lifestyle. The villagers (and the housing project residents less so) were not always forthcoming on questions about the house as it is a form of wealth and hence a closely guarded secret. Being a woman created opportunities for the writer to enter courtyards in the village or houses in the colony, and to talk to women of the family. But it proved to be a handicap as well; while it was easier to gain entry into the homes, spontaneous conversations with male informants which can be struck up in public places such as tea-stalls, market place, etc., were restricted.

The perception of the researcher by the informants is very important because upon that depends the information that they will share with him/her. The researcher's role is often confused with that of others in a traditional society. The informants may have trouble dissociating the researcher from his/her expected role in society when he/she is part of the same culture. He/she may be regarded as a member of an institution which is viewed with hostility and suspicion. In the village, the writer was suspected of being an employee of the local housing agency or the health facility, with both of whom the villagers have had unpleasant experiences. Though the climate of suspicion persisted in the housing colony (in many cases due to the unauthorised alterations made in the dwelling), there were fewer misunderstandings regarding her role as a researcher. While being a member of the same society is advantageous in some regards such as *a priori* understanding of norms, rules of conduct, and social class, being a foreigner may work to a researcher's advantage, particularly if he/she is perceived as a disinterested student of society.

SOCIAL CONSEQUENCES OF THIS PLANNING APPROACH

Since urban housing is a vehicle of social change, a post-occupancy evaluation of a colony gives a useful insight into the connections between social and physical change. As Franck (1985) points out, all designed environments result in environmental changes, i.e. they are purposeful modifications of our physical surroundings. This points to the need to examine the socio-cultural consequences of planned interventions so as to be able to judge their impact. For the housing colony residents (many of whom are rural migrants), as well as for the fringe village inhabitants caught in the process of expansion of the city through housing projects, social change is part of a larger process of urbanisation which has far-reaching effects on family structure, social networks, gender roles, and educational and economic opportunities. This can be demonstrated in a comparison between courtyard dwellings in traditional settlements in Northern India and the housing colony.

Residence in a housing colony implies social mobility on the part of families – they become owners of a house in a 'modern' (as opposed to 'traditional', 'congested', 'dirty') part of the city. Home possession becomes an indicator of occupational and social status, and the house is seen as an economic unit. The front room or the first floor may be rented out in order to recover the cost of building or provide an additional source of income. There are also absentee landlords who rent out the entire unit. One of the effects of state-provided housing is that there is a change in people's expectation of a house. The desired model becomes the middle-class, capital-intensive house built to last many years. The labour-intensive, high maintenance, temporary kind of housing is not what the residents want: it is associated with rural and squatter communities. There is also a greater dependence on public institutions for improvement in the quality of housing and the environment. In spite of the many complaints about the upkeep of streets and open spaces, there is no organised, co-operative effort either to pressurise the housing board to provide better maintenance or improve neighbourhood amenities.

The meaning attached to the house is a product of social relations. In the fluid social world of the housing colony, where the social relations have not been cultivated over several generations as in a traditional neighbourhood, identity and status in new groups are affirmed through status markers symbolised in the house and the goods that it contains such as television, refrigerator, etc. However, the cultural use of space (dictated by the sacred and polluting nature of the functions carried out in it) is not forsaken easily. In Indiranagar, 99 per cent of the households expressed their belief in the need for separation of living quarters from

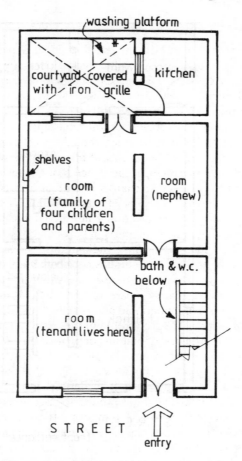

Figure 2.2 Plan of a typical house as altered and extended by the residents, Indiranagar

cooking and cleaning areas.[4] The core consists of a single room with an attached kitchen, WC, and bathing space. The survey revealed that 25 households shifted the kitchen to the back of the house in the space provided for the courtyard; 14 households constructed the WC near the entrance (see Figures 2.2 and 2.3). Rooms are seen as suitable for certain functions, open space in the form of a courtyard, and a lawn in front (primarily for display), are seen as necessary. Bathing space and toilet are preferred to be 'separate' from the living areas.

There is no consensus in the sociological literature about rural areas being associated with extended or joint family household type and urban

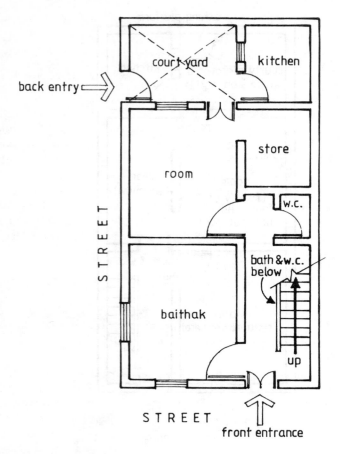

Figure 2.3 Plan of a corner house as altered and extended by the residents, Indiranagar

areas with nuclear family households. However, in the two areas of Lucknow under study, there seems to be a correlation between the village courtyard housing and extended family households, and the new housing and nuclear or modified nuclear family households.[5] Forty-one households surveyed in Indiranagar were nuclear families, five had a modified nuclear family structure, and only four were extended. The small size of the EWS house makes it particulary unsuitable for any but the nuclear family household. This has wide implications on a number of issues, including the use of space by household members within the dwelling, and development of new modes of privacy.

There is a close congruence between courtyard dwellings in their cluster arrangement and the joint family household based on kinship organisation and patrilocality. The courtyard house with its communal living spaces is particularly suited to large extended family households. Its form is adaptable to changes taking place in the life cycle of the family. Such changes include a nuclear family household growing into a joint family household or a joint family household breaking into smaller units of nuclear family household. In the courtyard house, the building takes place from outside in; rooms are added around the shrinking courtyard. When the household divides, the courtyard may be divided and the rooms may be claimed by individual households.

The position of an individual in the joint family household is determined by age, sex and generational status. Observance of authority entails avoidance-respect behaviour on the part of members lower in the hierarchy. The space in a courtyard dwelling is structured in order to ensure a smooth execution of this avoidance-respect behaviour through separation of male–female, and public–private quarters (see Figure 2.4). The courtyard dwelling is inward-looking and the private open space therein is for the benefit of the female members of the family who, in principle, do not use the public open areas. Any breakdown in the hierarchical social structure which characterises a joint family household results in diminishing use of rules of precedence as well as avoidance behaviour in the use of space.

In an EWS house in Indiranagar, there is no provision for expansion beyond the front room and the first floor. Even if there is maximum development, there is still not sufficient room for two households to live. The segregation between male and female spheres was found to be lower in Indiranagar housing colony than in the courtyard housing form organised on the principle of gender segregation in space. In the housing colony, there is a less sharp division between inside and outside and back and front due to the size and design of the dwelling as back-to-back row housing, see Figure 2.5). The dwelling is also more outward looking and is closely related to the street with none of the transitional areas between private and public realms which characterise a courtyard house. This, together with changes in household structure (from joint to nuclear), encourages a greater use of public areas by women.

While the structure of social relations in a traditional neighbourhood shapes the housing pattern and settlement form, in the housing colony the social structure was found to have been adjusted to the physical pattern already built. The neighbourhoods in the village or the inner-city are kinship clusters and display a close congruence between the spatial distribution of the houses and the kinship ties. Within a caste

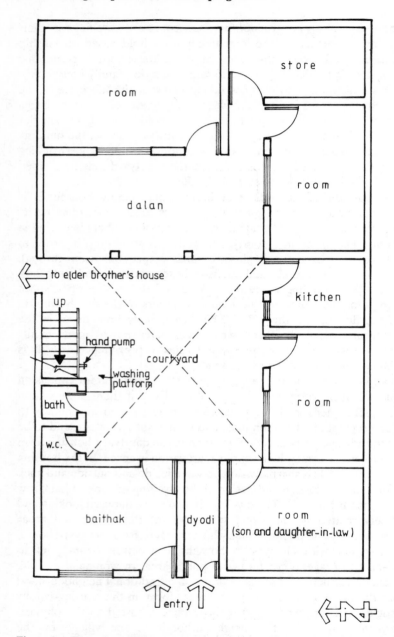

Figure 2.4 Plan of a traditional courtyard dwelling, Lucknow

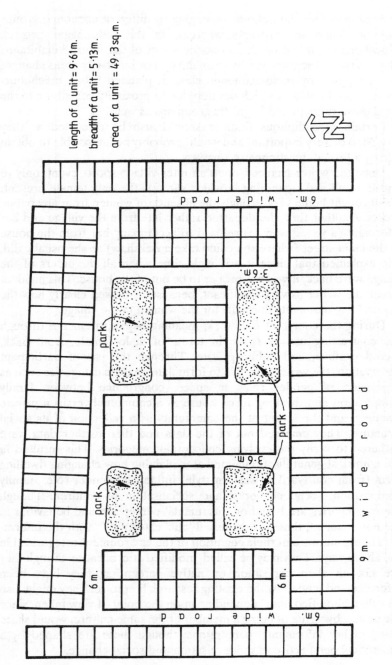

length of a unit = 9·61 m.

breadth of a unit = 5·13 m.

area of a unit = 49·3 sq.m.

9 m. wide road

6 m. wide road

6 m. wide road

6 m.

6 m.

3·6 m.

3·6 m.

park

park

park

park

N

Figure 2.5 Layout plan of a sector in the housing colony, Indiranagar

or kinship cluster, households belonging to different economic groups may be found in proximity whereas, in the Indiranagar project, allocations are based on the economic status of the future inhabitants. Thus territorial segregation by caste in rural or inner-city areas changes to segregation by socio-economic class in planned urban neighbourhoods. Social contacts which develop due to proximity contribute to the breakdown in caste and linguistic groupings as well.

Certain serendipitous findings should also be discussed as they proved to be very important and would probably be impossible to obtain during a formal questionnaire survey.

Once the writer chanced upon an inter-village sports event, only to find it an all-male gathering in which she was the only female present. This brought home very strongly the absence of women from any public events. Another time she departed rather late from the village and her informant (a young woman) refused to accompany her from the house to the main street (where she could take a rickshaw) as she usually did. She explained that since it was night-time when all the elders of the village were back, it was not proper to be out of the house. This incident struck the writer as very significant, because it showed clearly how the nature of spaces shifts diurnally for the women in the village.

During the course of fieldwork, meaning began to emerge through the contrasting use of space in terms of male–female, front–back, sacred–profane, and public–private. Through the use of participant observation, the writer was able to formulate analytic categories such as expression of gender roles in space, congruence between family composition and dwelling form, house as a communal versus a private dwelling, and the layout of the community as a reflection of its social structure. They emerged out of the data and then further data were gathered to verify the evolving conceptual categories. This resulted in the empirical generalisation aimed at explaining how changing dwelling form (from courtyard to western-style) influenced gender roles, family composition, social networks, and socialisation of children. Though these categories are based on informants' perception and behaviour in the environment, these are modified 'etic' categories[6], with the informants being unaware of their cognition of the environment categorised in the above ways. They may be called 'modified etic' because, though they are not an alien imposition on native perception, they have been informed and guided by the existing research literature. They might also be called 'modified emic' depending upon the point of view taken, since the writer, by virtue of being a member of the same culture, would share many of her informants' perceptions; though these are shaped by a theoretical perspective of modernisation and social change.

CONCLUSIONS

The above case study illustrates the shifts in attitudes towards housing which are taking place as a result of urbanisation and institutional provision of housing. In the discussions on state aided and built low income housing in the developing world, the impact of housing on social and cultural parameters of society is generally lacking. While it is recognised that there are no universal solutions to the housing problems in the developing world, systematic studies of how housing schemes perform in widely differing economic and cultural contexts are needed. When government housing projects accommodate rural migrants, they also serve to socialise the residents into urban ways of life. Thus the built environment contributes to the larger process of social change occurring as a result of urbanisation. Qualitative research methods are useful in post-occupancy evaluation of how residents alter and add to the urban house, the time when they undertake this building activity, and the stage in their life cycle when they find resources to do so; all these give clues as to how residents adapt to state-built housing. In state-aided housing, the role of the institution and the planning model used should also be studied as they introduce constraints as much as they facilitate self-help. The participant observation method will be particularly appropriate in cases where housing solutions have been put into effect in the interest of speedy provision without regard to the regional vernacular or the life-styles of the residents.

A research method which concentrates on the view of the user instead of just the interests of the planning authority requires flexibility, such as that provided by participant observation. A survey may have yielded data on the size of the family, but would have thrown little light on the shifting dynamics of interpersonal behaviour as the family composition changes from joint to nuclear. Similarly, the changing meaning attached to the house, as a rural traditional society moves towards urbanisation, could not have been captured with a standardised questionnaire. An open-ended case study approach, using participant observation, provides valuable insights into social and cultural factors in housing which may have been ignored in an evaluation which depended only upon surveys.

NOTES

1 The suburbanisation of the stable working class is similar to the process in the cities of the developed world, only it is state-aided in a developing country such as India.

2 The design is derived from the 'bungalow' type (an example of colonial architecture introduced by the British in India) consisting of single-purpose

rooms with attached bath and WC, and set in open space rather than containing open space within the building structure as in a courtyard. The same model is followed in all four types of housing; high, middle and low income group and economically weaker section (EWS); though there are variations in terms of size, layout and other amenities.

3   This view is shared by the architects in the Housing Board, as an interview revealed.

4   This concept is based upon the cultural attitude that functions carried out in space not only invest it with their sacred or polluting character but, since pollution is contagious, contiguous spaces may also be invested with it, necessitating the separation of the kitchen and the living quarters from the WC.

5   The Indian joint family consists of two or more married couples between whom there is a sibling or a parent–child bond. It may be supplemented by a widowed parent or an unmarried sibling. It is patrilineal in descent, patrilocal in residence, and patriarchal in authority. A nuclear family can be defined as a couple with or without unmarried children. A modified or supplemented nuclear family is one which contains a nuclear family plus one or more unmarried, separated or widowed relatives or parents (Kolenda, 1968).

6   'Emic' and 'etic' are anthropological terms referring, respectively, to making an effort to understand how things are seen and experienced from within, and to recording and analysing behaviour from the viewpoint of an outside observer (Barnouv, 1979: 12).

3 Cultural change analysis

Core concepts of housing for the Tswana

Amos Rapoport and Graeme Hardie

INTRODUCTION

This chapter consists of two major parts: a general approach or methodology for studying built environments in developing countries and an application of the specific method in a field study in Southern Africa.

The objective is to identify and differentiate between what are called core elements of a traditional environment and peripheral elements (those disappearing or being replaced by new, highly valued elements) and to create supportive environments incorporating both.

The design of environments for developing countries is particularly important because the problems are acute, and in urgent need of better solutions. The importance of such design is not only intrinsic, however, but also extrinsic; it is as a point of entry into broader issues and an exemplar for design generally and how it should be carried out (Rapoport 1983d: 240–54).

Design is seen also as a responsible attempt to help provide settings appropriate for specific groups of people; as a problem-solving activity which must be based on an understanding of environment-behaviour relations (EBR). Conceivably, a designer might design an environment that he intensely disliked if it were appropriate and supportive for the group in question. The designer's satisfaction comes from a problem understood, analysed and solved.

In order to be useful, an EBR approach must be based on theory; on a cogent, coherent overall conceptual framework. EBR is more than a tool to aid in programming and design; it needs to be seen as an emerging new theory of design.

The purpose of theory is to set goals and objectives and to provide criteria for making choices among alternatives. The purpose of such criteria is to guide the answer to the question: what should be done and why? The question of how it should be done deals with implementation, with the various constraining and enabling variables such as economics, politics, structure, materials, site conditions and the like; while important, these are modifying factors and are not considered in this chapter.

The design process can be visualised as an attempt to create 'better' environments. Evaluating the success of such attempts is essentially a two-stage process: the first ascertaining whether objectives have been met and the second whether these objectives are valid? In this view, EBR theory is about the nature of objectives. If design is about creating 'better' environments then theory helps decide what is better, for whom, under what circumstances, why, how one knows it is better, and so on.

The theory on which design needs to be based requires research on the three basic questions of EBR: on the bio-social-psychological and cultural characteristics of people that should shape the kinds of settings to be created, on how environments affect people, and on the mechanisms linking people and environments (Rapoport, 1977; 1983b). In any given case the specifics of the group and the situation require environment-behaviour research. In the case of developing countries, designing 'western' environments or trying to copy forms should be avoided. It can be shown (Rapoport 1983a: 251–3) that copying (or 'design by imitation') in general is unlikely to be accepted and that, more specifically, very serious mistakes are likely if shapes and 'hardware' are copied unless the type of research being described is carried out.

The approach to be discussed was developed on the basis of theoretical considerations and intuitive arguments for the importance of culture- and group-specific design by analysing traditional environments (for example, Rapoport, 1967; 1969a, b; 1975), and their implications for design in developing countries (Rapoport, 1979; 1980b).

THE MODEL AND CONCEPTS

In the present context 'development' and 'modernisation' are seen as equivalent to culture change. More specifically, one is dealing with a form of acculturation since the changes clearly seem to be due, in major part, to intercultural processes (contact, interaction or conflict) between the local traditional culture and the modern Western culture (Rapoport, 1989a). Complete rejection of either the new or the traditional components and characteristics of either built environments or cultural attributes are seen as equally unlikely and undesirable.[1] In any given case, the result of culture change on the built environments will be (or should be) some form of syncretism or synthesis between certain core elements of the traditional culture and environment, and important and highly valued elements of the new (Rapoport; 1983a: 255). In this, modern imagery may be as important as some traditional elements; in fact, the syncretism may be precisely between these two. Moreover, traditional elements may need to be provided almost surreptitiously by being given modern imagery or they may be rejected (Rapoport, 1983a: 251–3). One reason for this is that wants are as, or more, important than needs or (more correctly) needs are often latent rather than instrumental; the image of modernity may be crucial (Rapoport, 1983a and 1990a).

The complete replacement of the traditional by the modern, is usually undesirable. Elements from both are necessary if built environments are to be both wanted and supportive, although supportiveness usually involves traditional social units and institutions and suitable settings for those. These traditional elements will, however, only become acceptable if desirable images and meanings of modernity are being communicated and many of its undoubted instrumental advantages provided (in comfort, health, reduced maintenance, etc.).

Appropriate planning and design are taken to be culture-specific, i.e. group specific, so that relevant groups need to be identified (Rapoport, 1980a; 1985). Social and cultural variables are critical in helping to define the nature of relevant groups, to describe their life-styles, values, preferences, and the nature of 'good' or 'better' settings for them.

Given the great cultural diversity of user groups, one cannot generalise even for any one country. Even single cities clearly contain a great variety of subgroups that are potentially relevant for planning and design (Rapoport, 1977; 1988). Neither are tribal, racial or ethnic groups necessarily relevant, i.e. equivalent to cultural groups. For example, such groups classified in the USA as 'middle-class white ethnics' in a city like New York are not useful, since Italian, Irish and Jewish subgroups (to mention just three) are very different (Kornblum and Beshers, 1989). In any case, and this is the important point, one does not know whether the groups used are relevant for design.

There is little research on how groups are to be defined, and no explicit methods are available. At this point it is best to base choices on various aspects of culture (Rapoport, 1980a; 1986b; 1990c: fn4.). The definition can be based on the literature, general knowledge, the researcher's own knowledge, etc. – and treated as a hypothesis.

Once relevant groups are identified, two contextual variables need to be taken into account; and cannot be unless the group is properly identified. These are hypotheses based on previous work.

The first contextual variable concerns what one might call conceptual or cognitive distance, i.e. the extent of the differences in life-style, social organisation, values, behaviour and built environment between the traditional and the modern culture. The greater this is, the more difficult, disruptive, and potentially destructive is change, the more modulation of rates of change is needed, and the more supportive does the environment need to be. For example, problems of urbanisation are likely to be much greater for hunter-gatherers than for villagers or urbanites; i.e. criticality increases with cognitive distance (Rapoport, 1969a; 1978a, b; and below). This is partly because so many aspects of culture and built form need to be changed at once.

The second contextual variable is the rate of change. Problems of development and modernisation, therefore, are less related to change itself than to radical, abrupt, and excessively rapid culture change. In general, the faster the rate of change the more disruptive and potentially destructive it is. Again, criticality increases, and with it the need to modulate rates of change by providing supportive environments, as change becomes faster.

When both contextual variables apply, criticality increases even more and built environments play a more important role (Rapoport, 1978a, b; 1983b). Supportive environments reduce or eliminate stress by modulating the amount and rate of changes, thus providing (or 'buying') time for creative synthesis to occur. While all environments need to be supportive, this is particularly the case in developing countries. The purpose of the methodology being developed and discussed is to discover what would be supportive environments in any given case.

The notion of a supportive environment follows from the second basic question of EBR; the effect of environment on people (Rapoport, 1983b). While environments are not determining, they do have effects on people, particularly in conditions of reduced competence (such as physical disability, rapid culture change, etc.) or heightened criticality. Thus environments which are merely inhibiting under conditions of low criticality may become negatively determining, and even destructive in extreme cases, under conditions of high criticality; as in the case of developing countries (Rapoport, 1977: 3, 259, 264; 1983a, b). In some cases environments may need to be so highly supportive that they may be considered to be 'prosthetic' (a term originally developed by M. Powell Lawton in the context of environmental gerontology). It is then useful to ask three further questions about supportive environments:

1   What is being supported?
2   By what is it being supported?
3   What mechanisms are involved in its support?

What is being supported is typically a set of particular, important activities, components and institutions of the culture. In this connection the concept of mediating structures is important[2] (Berger, 1981, 1983; Badura, 1986), and this is more critical in developing countries. In that sense appropriate settings provide support 'indirectly'; they make more likely the survival of certain social units or networks, behaviour, activities, institutions and the like; it is these that are ultimately supportive. This is an example of what one could call indirect effects of environment on behaviour (Rapoport, 1983b; 1990a).

In the design of built environments, these social and cultural structures and institutions need to be linked to physical settings, i.e. fixed-feature and semi-fixed feature elements, including not only dwellings and other specific settings (which need to be discovered) but the system of settings (the nature and extent of which also needs to be discovered, not assumed or, worse, ignored) (Rapoport, 1977; 1980c; 1986a; 1990c; cf. Vayda, 1983). This linkage has been neglected, indeed ignored, by sociologists such as Berger and by development economists, applied anthropologists and the like.

It might be thought that research intended to discover all these matters can be avoided, and the concerns disappear, by relying on the provision only of infrastructure frameworks within which users construct their own dwellings (e.g. 'sites and services'). But merely leaving the building of dwellings to inhabitants is insufficient to make them congruent with life-style and other critical elements of the culture; in effect the problem is only shifted so that similar mistakes are likely to be made in a different context and at a larger scale (Rapoport, 1980b). Decisions taken about infrastructure layout at these larger scales may interfere with the layout of house groups or compounds, micro-neighbourhoods or communities, and can distort or even block the ability of inhabitants to achieve the desired spatial organisation, privacy mechanisms, front–back relationships, and other critical relationships pertaining to the dwelling itself. Even more important, seen in terms of activity systems and social groups and networks, the dwelling is an inadequate unit of analysis. The whole system of settings (or house-settlement system) needs to be considered (Rapoport, 1969a; 1977; 1980c; 1986a; 1989b; 1990c). Important activities may occur in various parts of the system of settings (men's houses, stables, outdoor space) or in other, often counter-intuitive, parts of the system. By ignoring the culture-specific nature of the overall layout of communities, one is, in effect, ignoring the nature of the appropriate system of settings.

There thus remains an inescapable need to identify the elements of supportive environments; what is important cannot be guessed or assumed but must be discovered; it could be anywhere in the system of settings (see below).[3]

THE PROBLEM OF SUPPORTIVENESS

The answers to the three questions about supportiveness are related to bio-social, psychological and cultural variables. Here the last are the most directly involved because they lead to the specifics of what is being supported, by what, and how, for any given group. One needs to discover the group's important characteristics and how these interact with various elements of the built environment. Even when the group has been identified, it is still far from a simple matter to decide what would be an appropriate, i.e. supportive, environment. Not only is environmental quality not always intuitively clear, it may even be counter-intuitive. A seemingly obvious approach may be highly ethnocentric and hence wrong; apparent improvements may have negative consequences; apparently undesirable conditions may be highly supportive for particular groups. For example, the provision of walls, replacement of earth floors by concrete, provision of electric lighting or running water may all have highly damaging consequences (Rapoport, 1978c).

It is necessary to identify the cultural specifics involved, to establish relations between crucial aspects of culture and of settings supportive of those.[4] It seems intuitively likely that, in any given case, certain aspects of culture will play a more central role; those need to be discovered. Generally, however, what are being supported, by various settings forming a system, are social units, institutions and the like (kinship and other groups, rituals, language, food habits, activity systems, etc.).

It is, therefore, necessary to be able to identify those elements of the culture which are essential to the identity of the group (both to itself and to others) and, hence, to its continuity. These elements together can be called the culture core. Then, those elements of the built environment which are supportive of the culture core should be identified and their important characteristics described. It is essential to understand how the elements of the culture core interact with various components of the built environment: which elements are supported, by what, and how? The objective is to specify the environmental quality characteristics of the profile and their relationship to the life-style of the group (cf. Rapoport 1977; 1980a; 1990d).

Ideally, and in greatly simplified form, it has been found that, for any group, four sets of things need to be identified.

1 *The relevant critical, central or core social units of the group and their role in the culture.* These may be castes; kin, age, ethnic, religious, initiation, linguistic, and other groups; social networks, etc. Associated with all these are activity systems, life-styles, values, etc.

2 *The corresponding physical units at different scales.* This reinforces the importance of identifying the systems of settings in which given systems of activities take place because appropriate housing form is closely related to other settings and the larger settlement system. As will be seen later, settlement form may be more important than dwelling form, or vice versa. More typically, elements of both, in specific relationships, comprise a supportive system. The specific components, relationships among them, and how these support activity systems, groups and institutions in different and highly culture-specific ways, need to be discovered. Semi-fixed elements (furnishings, artifacts and possessions) are also important; in many cases an increase in the number and value of objects owned has major impacts on both culture and settings (Rapoport 1989a; cf. Yellen, 1985).

3 *The units of social integration or interaction for the group in question and with other groups.* In the latter case, questions of 'neutrality' (Rapoport, 1977; 1980a; 1982), location and so on can become very important. Again, all these matters need to be discovered, and may not be self-evident.

4 *The institutions of the group, the highly specific ways in which certain common economic, recreational, ritual, governing, and other activities are carried out and, consequently, the highly specific settings for these.*

The essential step in design is to specify what should be done and why. This requires the identification of relationships between culture-core and supportive elements of the traditional built environment, of important new elements, and how these can be derived, logically and explicitly, from the specifics of the group.

It should be noted that, in general, data on social and cultural characteristics and changes in them; in values, behaviour and institutions, etc.; tend to be much more readily obtainable from the literature than comparable data on built environments; systems of settings and semi-fixed feature elements, etc. The problem is, thus, to discover especially the latter type of information.

THE APPROACH OR GENERAL METHODOLOGY

The process of culture change and associated changes to the built environment is dynamic, so that the problem is to identify persisting traditional elements, as well as disappearing, changing, or new elements. Almost by definition, any approach that attempts to obtain such information will be longitudinal or historical, even if it is not explicitly so (e.g. Hardie, 1980; Yellen, 1985; Hublin, 1989). In that sense, this approach highlights the need for longitudinal research in environment-behaviour research generally (Rapoport, 1986b) and the importance of historical data for generalisation and theory development in environment-behaviour studies (Rapoport, 1990b). It is important to bear this in mind; the method is essentially historical in that it tries to establish a baseline of the traditional environment (and culture) and then, through a sequence of built environments over time, to identify persisting core traditional elements, and changing, disappearing and newly introduced elements. It seeks to discover, in a highly specific context, the changes and dynamics of the situation, so that what is being given up, modified, retained or adapted, provides clues as to the relative importance of elements, and inferences about why they are important. This relative importance may enable the identification of the culture core and those elements of the built environment associated with it, which this study emphasises.

This first step identifies those elements that are changing slowly or not at all. Although distortions due to political, economic and other constraints occur, the changes found can frequently be relied upon to indicate priorities and give clues as to the relative importance of elements. For example, among the Navaho (and also among the Bedouin) house forms seem to change relatively rapidly. Modern houses, modern materials, and modern appliances are desired, although they may continue to be used in traditional ways (for example, Kent, 1984) or retain traditional space organisation (as among the Bedouin). In the case of these two groups, however, the settlement pattern seems to change much less rapidly and, if changed, seems to cause resistance and negative consequences. The traditional settlement form seems to be central, with houses arranged at low density, or at low perceived density (Rapoport, 1975b; 1977) with proximate houses belonging to kin, leading to the retention of particular social groupings related to housing clusters separated by space (Sadalla *et al.*, 1977; Rapoport, 1978b; 1983a: 265–6). In these two cases, dwellings have changed, are changing or are required to change rapidly, whereas the settlement pattern tends to retain its traditional form and to resist change: it represents the core environmental elements.

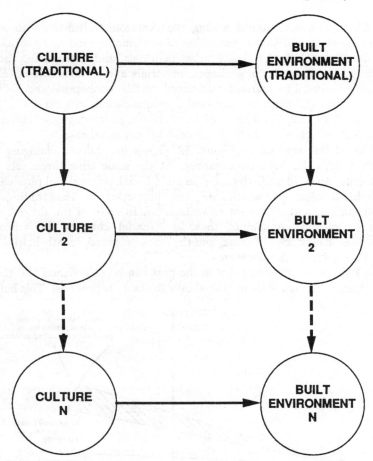

Figure 3.1 Culture changes in tandem with the built environment
Source: Based on Rapoport (1983a: fig. 4, p. 259).

In some cases, however, house forms have changed less and seem to be much more central. Moreover, these constancies in house form are not apparent without analysis (Hardie, 1980 and below). Among the Tswana (in certain areas of Botswana), settlement forms changed easily and very early. Thus at one level, the case of the Tswana is the exact reverse of the Navaho and the Bedouin. It was, in fact, this apparent contradiction that led to the development of the present method.

At another level, traditional elements, such as organisation, persist even in transformed portions of the environment; but only detailed analysis will reveal this. Not only is this the case with the Tswana dwelling, it is also found among the Amazonian Indians who are retaining the traditional organisation of settlements (and some aspects of dwellings) although using new materials and introducing new settings. For example, new shapes, materials and even activities (e.g. soccer) are used to recreate traditional spatial, conceptual and social organisation (Cristina Sa, personal communication, August 1982; cf. Rapoport, 1977: 8, 11). The result is, of course, a high degree of cultural specificity which needs to be discovered for any given case.

The global analysis in Figure 3.1 shows the culture changing in tandem with the built environment. At the same time three sets of elements can be identified: core elements (social, cultural and physical); peripheral elements which are less important in the traditional environment, and important new elements. In spite of the distortions due to various constraints, these differential changes do indicate priorities. Knowledge of these, and the clues provided, greatly help with predicting future developments.

A baseline or starting point in the past can be established and then the changes, or lack of them, traced over the more recent past. This helps

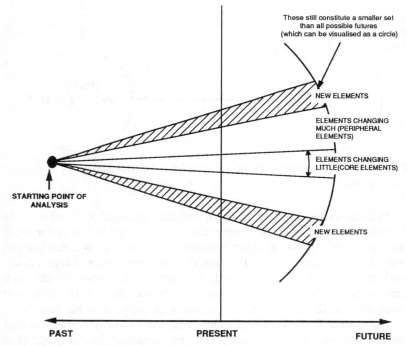

Figure 3.2 Tracing the changes over time
Source: Based on Rapoport (1983a: fig. 5, p. 261).

to project the likely extent of the change in peripheral elements. This may still be large, but is smaller than what would be faced without this type of analysis; one is dealing with the arc of a circle rather than a full circle. As for the core, this is very much smaller, possibly almost insignificant over the next 15–20 years or even longer. Given the notorious difficulty of forecasting the future, even such crude indications are very useful.

It is important to realise that analysis follows the process shown in Figure 3.3. The starting point at one end is the traditional environment which has certain characteristics congruent with components of the culture. In trying to identify and understand changes in it, two things are found. First, the changed environment has a reduced set of characteristics congruent with the core elements of the culture; this is what remains of the traditional environment. Second, new characteristics both in the environment and the culture can be found. These two sets come together, resulting in the synthesis or syncretism discussed earlier.

In more detail, and this is the key element of the method being described, the analysis proceeds from two directions, as shown in Figure 3.4.

Starting with the left-hand side of Figure 3.4, the analysis begins by trying to identify the characteristics of the traditional environment as a baseline. Archaeological data can sometimes be useful. What was the pre-contact situation? During our fieldwork we were given aerial photographs of newly discovered archaeological remains of early Tswana settlements which appear not to have been published or even studied. What are the first descriptions from travellers or ethnographers? Hardie (1980) uses both pictorial and written material to establish the nature of Tswana dwellings and settlements in 1801.

Next, modified traditional environments in the most remote villages or settlements are examined to determine what were the first elements to change. Then, spontaneous settlements in small, provincial towns, large cities or the capital are examined in succession. Within a given location, recent environments might be examined rather than those longer established. (In all cases group specificity would be considered, at least ideally, not only in environmental elements but also in activity systems, life-style, institutions, social groups, etc.)

The number of steps to be used will depend on various circumstances, but in essence they follow from the traditional to the most modified. The right-hand side of Figure 3.4 begins with the image of modernity; perhaps what television programmes and magazines (the source of much imagery) are showing. Next, the private housing of elites, judges, ministers and senior civil servants, are examined because that is likely to

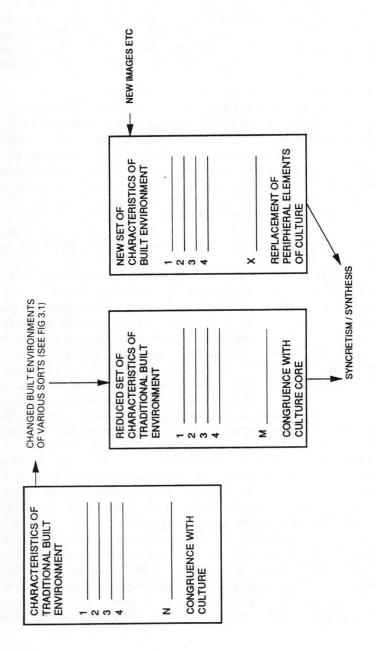

Figure 3.3 Identifying the characteristics of traditional, changed and new environments

Source: Based on Rapoport (1983a: fig. 6, p. 263).

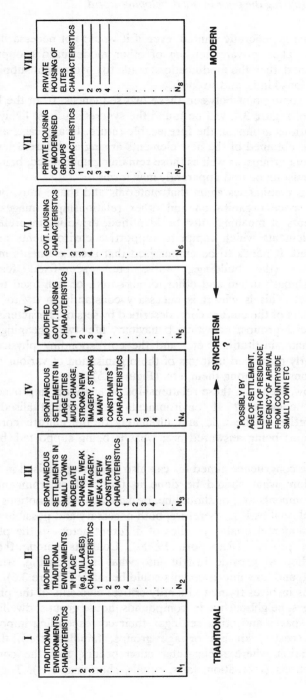

Figure 3.4 Identifying the characteristics of environments ranging from traditional to modern
Source: Based on Rapoport (1983a: fig. 7, p. 262).

be what is generally wanted, even if it would not necessarily best meet needs. Then private housing of other modernised groups would be examined, then the modifications made to government supplied housing of various kinds, and so forth.

At some point between these two sequences, from the left and the right of Figure 3.4, will be found the syncretism most likely to develop for various groups in the foreseeable future. In any case, an indication will be obtained of the new elements arising at different levels and for different groups, as well as those remaining unchanged, being recreated or persisting behind apparently new forms.

This emphasises again that materials and shape must be separated from space organisation and other relationships; images and other elements of meaning must be identified, priorities in preferences and the elements which comprise supportive environments must be discovered. It needs to be emphasised that these elements may be fixed features (the buildings, walls, etc.), semi-fixed features (the 'furnishings'; urban and other, of all sorts), or non-fixed features (the people). This is why it is necessary constantly to try to identify the elements of the culture core, described by sets of characteristics related to social groups, activities, behaviour, life-style, meanings, cognitive schemata, institutions, etc. and the corresponding physical elements similarly described in terms of the organisation of various settings, the relationships among them, etc. (Figure 3.5).

How, then, can these relationships be analysed, improved and made more congruent? If environments are conceptualised as being supportive of life-style and ultimately of the culture core, then the questions being answered are: what is being supported, by what and how?

The congruence aimed for can also be conceptualised in terms of the question: what should be done in terms of environmental quality? Environments are, or should be, designed to match notions of environmental quality. It is, therefore, both necessary and possible to study the environmental quality profiles of different actors in the planning and design process (Rapoport, 1990d). Conflicts among them can be identified to provide insight into what is happening, what is going wrong, and how improvements could be made (Figure 3.6).

This involves trying to identify critical elements of the physical units (villages, neighbourhoods, compounds, house groups, dwellings, rooms, open spaces and other settings), their corresponding important social units (castes, kinship or age groups, families, etc.), the units of integration where social interaction occurs, and the corresponding institutions (recreation, fairs, rituals, shopping, etc.). Together these

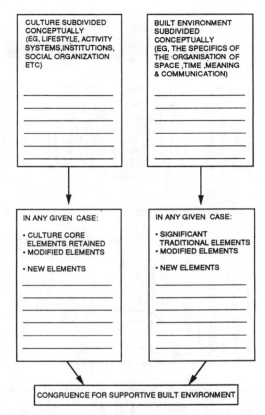

Figure 3.5 Elements of the culture core and corresponding physical elements
Source: Based on Rapoport (1983a: fig. 8, p. 263).

may begin to provide insights into appropriate relationships between activity systems (including their latent aspects) and cognitive schemata expressed in domains and systems of settings. These often have clear equivalents in traditional environments, so that urbanisation and its difficulties can be discussed in the light of both social and spatial elements. For example, in some areas, homogeneity of areas (such as neighbourhoods) is very important, in others less so; in some places ethnicity may be the basis for such homogeneity (e.g. New Guinea), in others not (e.g. the Philippines). In yet others, it may be tribe, religion, language, occupation or whatever; the bases for perceived homogeneity differ (Rapoport, 1977; 1980–81).

The analysis being described can help identify the consequences of such processes. In some cases there are ready-made spatial or physical

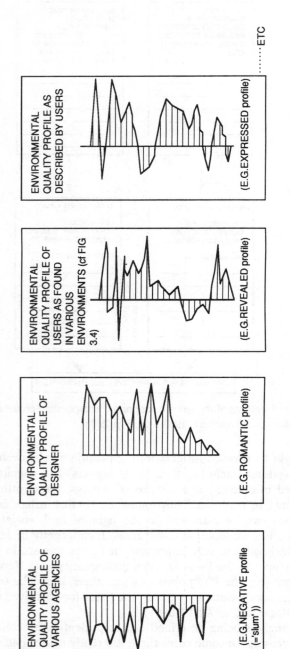

Figure 3.6 Environmental quality profiles from different perspectives
Source: Based on Rapoport (1983a: fig. 9, p. 263).

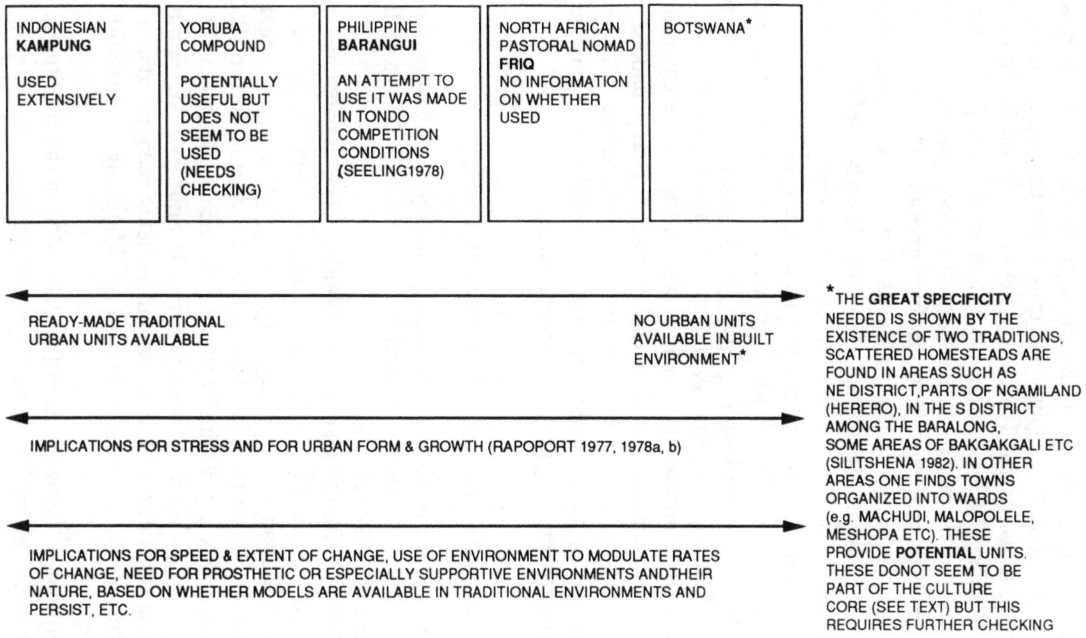

INDONESIAN KAMPUNG	YORUBA COMPOUND	PHILIPPINE BARANGUI	NORTH AFRICAN PASTORAL NOMAD FRIQ	BOTSWANA*
USED EXTENSIVELY	POTENTIALLY USEFUL BUT DOES NOT SEEM TO BE USED (NEEDS CHECKING)	AN ATTEMPT TO USE IT WAS MADE IN TONDO COMPETITION CONDITIONS (SEELING 1978)	NO INFORMATION ON WHETHER USED	

READY-MADE TRADITIONAL URBAN UNITS AVAILABLE ← → NO URBAN UNITS AVAILABLE IN BUILT ENVIRONMENT*

IMPLICATIONS FOR STRESS AND FOR URBAN FORM & GROWTH (RAPOPORT 1977, 1978a, b)

IMPLICATIONS FOR SPEED & EXTENT OF CHANGE, USE OF ENVIRONMENT TO MODULATE RATES OF CHANGE, NEED FOR PROSTHETIC OR ESPECIALLY SUPPORTIVE ENVIRONMENTS AND THEIR NATURE, BASED ON WHETHER MODELS ARE AVAILABLE IN TRADITIONAL ENVIRONMENTS AND PERSIST, ETC.

*THE GREAT SPECIFICITY NEEDED IS SHOWN BY THE EXISTENCE OF TWO TRADITIONS, SCATTERED HOMESTEADS ARE FOUND IN AREAS SUCH AS NE DISTRICT, PARTS OF NGAMILAND (HERERO), IN THE S DISTRICT AMONG THE BARALONG, SOME AREAS OF BAKGAKGALI ETC (SILITSHENA 1982). IN OTHER AREAS ONE FINDS TOWNS ORGANIZED INTO WARDS (e.g. MACHUDI, MALOPOLELE, MESHOPA ETC). THESE PROVIDE POTENTIAL UNITS. THESE DO NOT SEEM TO BE PART OF THE CULTURE CORE (SEE TEXT) BUT THIS REQUIRES FURTHER CHECKING

Figure 3.7 Extent to which traditional urban forms may be helpful for modern urban development
Source: Based on Rapoport (1983a: fig. 10, p. 264).

units, for example the Indonesian Kampung. As towns 'modernise' or develop, and the countryside becomes urbanised, the rural Kampungs are transformed into urban Kampungs, but the traditional, highly integrated social and physical unit tends to persist. Materials, amount of vegetation, density, etc., change, but the Kampungs continue to offer very useful, ready-made units to be used in the planning and design process.

A weaker form is found in the Barangui of the Philippines, which is more of a social unit; but physical equivalents can easily be developed (as the conditions for the Tondo scheme competition for Manila made clear; cf. Seeling, 1978). Others yet, such as the Friq of the pastoral nomads of North Africa have potential for spatial expression in sedentary and even urban situations (cf. Rapoport, 1978b). At the other extreme, traditional settlement patterns of widely scattered, individual homesteads provides no spatial elements directly useful for urbanisation; although potentially useful elements can still sometimes be derived through this form of analysis (see Figure 3.7).

This type of analysis can also predict problems and their potential severity. It allows prediction of the likely degree of stress and the problems likely to arise, thus giving opportunity to decide how best to modulate change and begin to specify what a supportive environment is likely to be. It seems likely that, in a culture where ready-made social and physical units exist and can be incorporated, there will be fewer problems than in a case where there is no tradition of urbanism, or no usable units; where cognitive distance is great.

A CASE STUDY: MMABATHO (MAFIKENG)

In order to test the method of producing a highly culture-specific design, a three-week pilot study was carried out among the Tswana living in Mmabatho (Mafikeng).

The Tswana differ from the other Southern Bantu tribal groupings in having lived for many centuries in consolidated settlements of as many as 5000 inhabitants. Mafikeng 'the place of stones', began as a traditional capital of the Hurutshe, a Tswana people, in the 1850s long before there was any White/Western influence in the town. There are few Bantu names used for major centres in Southern Africa but it says much about the nature of the place, and its being recognised as a Tswana capital town, that it is still known as Mafikeng; its traditions are respected.

Subsequent settlement has not interfered with the traditional capital but rather built alongside it. Both the British, who after Union in 1910 continued to administer their Protectorate of Bechuanaland (Botswana) from Mafikeng (although it was located outside of the Protectorate itself), and the South African Government developed housing around the town. The ubiquitous matchbox, four-roomed rental housing was built for the Black residents on a grid pattern. But the Government was unable to keep up with the population growth and so new residential areas developed around the town on land made available by the chief, according to tribal custom, with the people building their own houses.

More recently Mmabatho, yet another town built on the edge of Mafikeng, became the capital of the so-called independent state, Bophuthatswana, created by the South African government, the sovereignty of which has yet to be recognised by the international community. This resulted in new housing developments available for freehold purchase to all those who could afford the housing; mainly Government civil servants and professionals.

The relationship between the house and settlement, and cultural continuities and changes among the Tswana in Botswana, have been studied in detail (Hardie, 1980; 1982; 1985a; 1985b and 1986). However, it was felt that Mafikeng/Mmabatho was likely to be a particularly difficult case based on the particular political and economic conditions in South Africa and the effects these have on the group in question (Hardie and Hart, 1989). It was likely that individuals had worked in many parts of South Africa, including large metropolitan areas, had served in the army, had moved or been moved periodically and were essentially isolated from the cultural heartland and hence from important institutions, social units, and other aspects of culture. Thus, if the methodology and approach were able to 'work' and to provide some evidence, however slight, for the persistence of core elements of the built environment in this difficult case, then the approach should be far more useful in more 'typical' cases.

Before the specific methods used and the findings are presented, the usefulness of the method at the initial reconnaissance stage should be discussed briefly.

Simply through reading a tourist brochure and taking a quick drive around, the categories of environments used in the analysis (Figure 3.4) allow a researcher to make immediate sense of the town and its surroundings, and to divide it up in corresponding ways. Remarkably the full range of environments (except a traditional village) were found within this one town, and the village could be found within 30 km. Since getting oriented and making sense of places is often a most difficult and time- consuming matter, this was a major advantage of the method and supported the validity of the sequence of environments proposed and

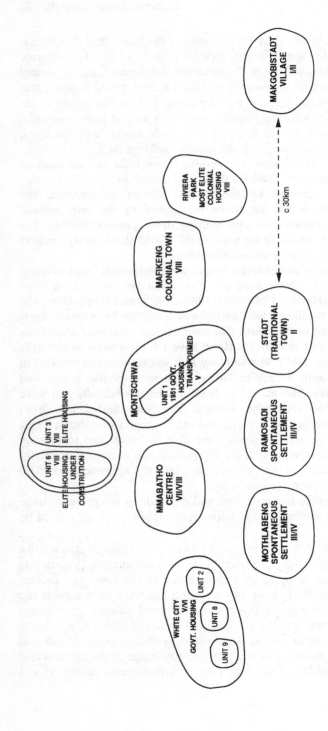

Figure 3.8 Mmabatho (Mafikeng) divided into each type of environment found in Figure 3.4

Note: Roman numerals refer to categories in figure 3.4

Note: Roman numerals refer to categories in Figure 3.4

MMABATHO/MAFIKENG

MAKGOBISTADT VILLAGE I/II

RIVIERA PARK MOST ELITE COLONIAL HOUSING VIII

MAFIKENG COLONIAL TOWN VIII

c 30km

STADT (TRADITIONAL TOWN) II

MONTSCHIWA
UNIT 1 1951 GOVT. HOUSING TRANSFORMED V

RAMOSADI SPONTANEOUS SETTLEMENT III/IV

UNIT 3 VIII ELITE HOUSING

UNIT 6 VIII ELITE HOUSING UNDER CONSTRUTION

MMABATHO CENTRE VII/VIII

MOTHLABENG SPONTANEOUS SETTLEMENT III/IV

WHITE CITY V/VI GOVT. HOUSING

UNIT 2

UNIT 8

UNIT 9

described in Figure 3.4. It also proved possible immediately to identify typical units and elements in each type of environment so as to photograph them and to decide which were to be recorded or studied in more detail. In effect, after arriving in the afternoon of one day it was possible to begin active research on the morning of the next day.

In Mmabatho (Mafikeng), it proved possible to consider housing occupied predominantly by one cultural group which covered a wide spectrum of house types, ranging in age and type as follows:

1 the rural farm homestead (category i in Figure 3.8);
2 the traditional capital of a head of state called 'de Stadt' (category ii);
3 the extensions and transformations of this town which initially closely follow the pattern of the original town (category iii, iv);
4 spontaneous areas which look more like any modern town (category iii and iv) than the traditional 'Stadt';
5 the government built housing which ignores all individual identity (category v);
6 the transformation of that housing which came about as the occupants became owners (after the Government made them available to the occupants for purchase);
7 new developer-built housing estates with a limited number of model house types (one of these estates even attempted to copy the formal forms of traditional settlement patterns but failed hopelessly) (category vi and vii);
8 and the individualised luxury housing of the elite (category viii).

In all, eight different residential settings were found. It was remarkable that all eight categories from traditional to modern in the original theoretical model should be found in one town. Thus, it proved an almost ideal testing ground for this investigation.

Methods available to compare area with area have been described in detail elsewhere (Hardie, 1989). These include: participant observation, photographic comparisons, detailed site drawings, projective techniques using models, in-depth interviews, and large questionnaire based surveys. Door-to-door interviews were felt to be the most effective method in this case. Ninety-two households in six of the eight segments identified were interviewed by three Tswana researchers using cluster sampling to select all households on a particular street or block. The rural homestead and the housing of the elite were sampled on an *ad hoc* individual basis. The interviews covered a wide range of topics which included describing the following: the features of the house and plot which were liked or disliked; the places used for socialising and eating at different times of the day; where householders shopped; societies, groups or associations to which they belonged; the inheritor of the house; what would be considered to be the dream house of the homeowner; and the household composition and residential history of the homeowner. A free-hand sketch was made of each house yard, noting the finish of the garden and yard; some which seemed typical were measured in more detail. Finally the interview also included a detailed inventory of all the household objects and furnishings.

ANALYSIS OF THE DATA[5]

A number of changes were immediately obvious. In the most traditional settlements the house had been incidental and was only used as a sleeping area whereas the yards were found to be highly articulated with spaces designated for particular activities. For example, there was a place for washing clothes and a separate enclosure for cooking; moreover, it was made clear by the residents that washing was never done in the cooking enclosure.

The use of the house and yard differs across the eight types, showing a fundamental change from the traditional pattern to that of the modern house. In each successive area the house becomes more and more dominant and also more articulated internally, whereas the yard becomes less articulated for household activities with the concept of a garden becoming dominant. However, where residents were building their own dwellings (in the elite areas) or transforming them (in government-built housing), it was found that the articulations of space exceeded those of most modern/western houses. The designation and specificity of spaces now moved indoors and, where possible, was highly elaborated. We found that rooms were specially designated for studying (for the school children separate from their bedrooms), or for playing the piano, or for TV viewing.

In the most traditional areas, there was a clear sense of what actions took place in the front yard, open to the public, and what happened in private at the back. This was a feature which did not change across categories. In fact, in the most modern houses the separation was made extremely clear in the internal arrangement of the house rather than the yard. There were rooms set aside for the visitors and there were also areas in which the family socialised, but visitors did not have easy access to them. Where developers had built open plan houses, or the design of the house required one to pass through a kitchen *en route* to the bedroom, this was always mentioned as a negative feature of the house for it made public what was held to be private.

A further example of this issue of privacy concerned the traditional separation of washing and defecating. Where this was not provided for in the modern houses, the residents gave most detailed reasons why these functions should not be together. The desire to maintain the separation between what was considered clean and unclean should be respected.

Traditionally the household slept according to age and sex. When the children were young, they slept in the same enclosure as their parents. As they reached puberty, the parents built special enclosures with the men often having their own individual space and the women staying together. These units were separated and in different parts of the yard, giving each a great sense of privacy. In the typical 'modern' three-bedroomed house all the bedrooms are located together. The negative comments in these homes often focused on the bedrooms and their proximity (with a resulting lack of privacy) or that the master bedroom was not large enough to allow the smaller children to sleep there as well. In one elite home, this problem was solved because the parents had a suite separated from the bedrooms of the children which were located together with their own facilities. Here the luxury 'modern' dwelling allowed for the desired separation noted in the past.

The inventories of household objects and furnishings proved to be most useful for they displayed a dramatic change in the approach to objects. Comparing the pictures on the walls, these changed from being a clutter of large black and white photographs of parents or the elderly residents themselves, photographs of the King and Queen of Great Britain, and illustrated biblical texts, to predominantly bare walls of the new developer-built homes with the occasional large poster of a rural scene in a far-away land. Occasionally the biblical texts appeared as well. The furnishings changed from being small and functional in the older homes to being large and bulky in the modern dwellings with TVs and electronic sound equipment much in evidence. The modern houses had few items which could be considered antiques; obviously modern items were considered best. What was obvious was the great pride taken in the furnishing of the homes whether old or new.

Communal social spaces where much time was spent talking and entertaining guests were another important feature which persisted, although the modern houses rarely allowed for this to take place outside with a view of the street. However, those houses built with covered porches, often with chairs on them, were seen positively. Where viewing to the street was not planned, the residents created their own solutions. In one instance in the elite area, we found that the residents who wanted to watch and participate in what was happening on the street sat under their opened garage door.

In the past, the boundary of the yard was clearly defined and ritually protected as it was believed that this protected the life of the people in the homestead from negative spiritual influences. This concept was maintained in all the areas; and the yards were clearly demarcated and kept immaculately clean and without weeds within the boundary, however large the plot.

In one modern development the plots had been laid out in a so-called traditional style with the houses located around dead-end streets. This meant that the plots were of variable shapes and sizes; it was not clear where the boundaries were and this required detailed surveying for them to be pegged. When the new occupants moved in, there were major problems and unrest. Psychologists who were consulted discovered that the residents' anxieties stemmed from nobody knowing where their boundaries were. Until they became clear, residents could not work on their yards as they wished to do, for it was important for them to have the yard clean and in so doing claim their property, but they feared to antagonise a neighbour by stepping over the boundary (a potentially dangerous action and one to be avoided). However, once the boundaries were established, the problems were resolved and within a short time highly decorative precast walls were erected.

Traditionally, boundaries were conceived as spiritual in nature. If a neighbour harboured jealous feelings this might affect a household's well-being, so each lived in a way which would not cause such feelings. Neighbours in the traditional areas might often have been kin of some kind, and so this belief in jealousy was particularly prevalent and often mentioned. However, in the modern developments where neighbours were unlikely to have had any previous social connection, the only mediating factor was that all the residents could afford the same amount of housing. This seemed to dissipate the fear of jealousy. Thus the economic segregation of the community by housing costs seems to have positive implications.

The landscaping of the yards was another transformation of note. In the traditional areas the yard was cleared and very often smeared and plastered with a cow dung and mud solution. There was no planting except for full grown trees which were both status symbols and, more importantly, gave shade to the social meeting places. The reasons given for no small plants were that they attract snakes which are seen as messengers of the ancestors and are not welcome. Over the range of areas the garden becomes a progressively more desired feature. In the areas under tribal control some planting is incorporated in the yard; most often in a controlled manner with stones, often painted white (a colour related to the belief in ancestral protection) to control the potentially disorderly element. As one moves to the more modern areas, the gardens become developed with a great many plants. However, the gardens seldom become organic in form but rather hold to strict geometric patterns with lawns being used in very defined ways. Quite often the yards were almost completely covered with cement allowing for only very small contained gardens. This approach contrasted dramatically with the freeflowing herbaceous borders of the white residents of the town.

There appeared to be a strong desire for flexibility. Where residents upgraded their 'matchbox' houses, an incredible degree of ingenuity was evident as the homeowners planned their houses in very complex forms. The developer-built houses, usually on small plots, allowed for little of this transformation, and the residents noted this as negative features of those houses. They felt trapped and unable to fulfil their desire to alter their houses with ease.

CONCLUSIONS

This chapter has described an approach for examining the complex issue of cultural change and its implications for designers. The method allows the analysis of the process of syncretism so that it is possible for the designer to operate in a way which is sensitive to the people, their values and traditions, and also to the changes they are incorporating in their world.

However, in this chapter, the effect of constraints on implementation are not considered (cf. Rapoport, 1983a: 267–268 and Figure 11). When the method is to be used in design, the core and new elements which are important can be derived as a hypothesis, and then other methods can be used to test whether these are still important. This can be done in a variety of ways; often modelling and projective techniques are most appropriate (Hardie, 1986; 1988; 1989; Hardie and Hart, 1985; Hardie, Hart and Theart, 1986).

From the few examples in this case study, it can be seen that this method provided a very useful means by which core elements of the Tswana culture, spatially and physically expressed, could be traced from one location to another and the resulting syncretism understood. This made it possible to distinguish those aspects of the house which were maintained as a result of past traditions and also to observe what had been adopted and incorporated from the so-called 'modern' or western designs and to discern where these adoptions caused stress and thus were not supportive to the occupants.

It could be argued that the traditional culture had been unusually transformed and manipulated because of the context of South Africa. However, it was found that core elements were still expressed in all areas – even among the elite. By comparing with Botswana (as an example of a more typical developing country), these continuities could be identified more clearly, and their strength and supportiveness could be reinforced in the modern settlement and house designs. Sadly this has not always been the case, because imported designers lacking sensitivity to the values of the people have used inappropriate models. It goes without saying that 'most' people want to be seen as 'modern' and thus have a house which looks 'modern'. This was made clear to the authors when in a squatter settlement a man, on completing a model of his future home, looked up and said: 'With a house like this I will never need to fear the white man'. The fact that the internal organisation of the house was different from most white man's houses was less important than what it 'looked' like (Hardie, 1980).

There has been a tendency for development agencies to decrease the plot size of urban dwellings from those used traditionally in the hope of increasing urban densities. It was found in this case study, and is supported by other work in Lesotho (undertaken by Hardie) and in Botswana, that this severely diminishes the homeowners' ability to manipulate their space and, consequently, also diminishes their satisfaction with that space. It has been found that, if given usable building areas, homeowners will use the land inventively and in a highly articulated manner. By allowing for tenants in the plan, they may actually obtain very high urban densities.

Finally, there are implications here for design more generally. The method and approach, the problems, the ideas, and the requirements are similar in 'western' locales; they are just more difficult to identify. Increasingly situations are found of cultural pluralism, of diverse groups which need different environmental qualities; a similar need for supportive environments to respond to situations of culture change. To reiterate, the discussion provides an entry point into much larger theoretical and design issues. While what is proposed will help greatly to improve design in developing countries, it goes well beyond that specific problem.

NOTES

1   Note that there is a literature in anthropology on the various scenarios and outcomes possible in such cases. This will not be discussed here.
2   This concept was developed at a much larger, societal, scale and in connection with economic and political concerns.

3 Examples of student work in the course based on this method show striking variations among case studies. In each case study highly specific elements are discovered as central and critical. For example, in the case of Bedouin, tombs of saints or sheikhs were found to be most important. In the case of Jamaica and of the Lapps in Northern Sweden, a regional analysis was necessary to understand how the system of settings at smaller scales operated (Rapoport, 1983a: 257–8, 265–6).

4 Note that 'culture' as such cannot be used either in analysis or in design, being both too general and too abstract. It needs to be dismantled, and one of us (Rapoport) has proposed a way of doing so, that leads along one axis to life-style and activity systems (including their latent aspects) and along the other to various social variables, such as roles, institutions, kinship and other groups, etc. (e.g. Rapoport, 1977; 1980a; 1986b; 1990c). In any given case one needs to discover the specifics of those.

5 This is only a very preliminary analysis: a complete analysis remains to be done.

4 Time series analysis
A longitudinal study of housing quality in Lusaka

Ann Schlyter

INTRODUCTION

Almost all research questions in the field of housing can benefit from being subject to a time perspective. Many researchers integrate past time into their analyses by using existing documentation or by asking respondents about the previous situation. In a longitudinal study the researcher either continuously observes a process over a long period of time, or makes systematic re-visits to an area for data collection.

This chapter presents a case study of a time series analysis of the quality of housing in a squatter settlement known as George in Lusaka, the capital of Zambia. Quality is always experienced by someone and the opinion of what is good quality might differ according to culture, class and gender. Research on quality of living space is, therefore, dealing with the users and their experiences, as well as the physical structures.

In George, a longitudinal study, including short field studies at about four year intervals, has been carried out over a period of twenty years.[1] Studies like this are unusual for obvious reasons. Few researchers have worked under conditions which allow them to maintain an interest in one settlement over such a long period.[2] The George study was designed to improve basic knowledge of housing processes and a focus on the quality of space was maintained through all phases. The intention of this chapter is to illustrate the potential of the longitudinal method in qualitative analysis of housing and settlements, using the George study as an example.

A REVIEW OF THE GEORGE STUDY

In order to illustrate how the longitudinal method can be adapted to changing planning and policy situations, this section will briefly review the George studies, putting each field study in its context. With approximately a quarter of its population living in towns at the end of the 1960s, Zambia was one of the most urbanised countries in Africa. Furthermore, Lusaka was growing rapidly, at about 13 per cent per year, and almost 40 per cent of the population was living in unauthorised settlements. The first study, in December 1968, concentrated on the effects of rapid urbanisation on housing conditions.

The choice of George as a squatter settlement to study was guided by several factors. In 1968, it was one of the largest squatter areas which had grown rapidly since independence in 1964. Located close to the heavy industrial area, George was likely to be a rather homogeneous working class area. The political situation there was stable compared with other settlements. In addition, some background data were available from a socio-medical study which had been carried out there by one of the team. Similar kinds of judgements frequently influence research processes, although they are often not admitted but covered over within more rational arguments. Twenty years later, the original reasons for the choice of settlement are of less interest than the fact that George is still fairly typical of those settlements in which more than half of the population of Lusaka live.

In the report of the first field study (Lundgren, Schlyter and Schlyter, 1969), a detailed investigation of physical structures and use of space was used, not only for recognising the qualities of housing in the squatter area, but also for identification of the problems. Statistics on migration rates and population increase, and government and local budgets, indicated that the squatter settlements were likely to grow rapidly. Following a user-oriented approach, comments from the inhabitants were compiled and conclusions drawn; that the area ought to be improved and legalised. To show the feasibility of such a policy, a draft redevelopment plan was worked out, including roads, water and services within the existing settlement.

The first investigation to a large degree determined the design of the subsequent field studies. The longitudinal method requires that the reinvestigations are carried out in the same or as similar a way as possible. When the study was initiated, it was not intended to be the first in a series. However, even during the fieldwork, the possibility of a restudy in order to assess the dynamics in the housing process was discussed.

Like other governments in the 1960s, the Zambian government regarded squatter areas as temporary failures which were to be demolished. There were, however, people within the planning authorities and in non-governmental organisations who gradually developed a case for upgrading and legalisation of existing squatter areas. The first report was found to be useful in this process as it was prepared by independent researchers based on an investigation of how the environment was formed and used.

In the years that followed, there was intense discussion about how upgrading could be carried out. The second field study, which was carried out in 1973, therefore aimed at making a contribution to this discussion. An understanding of the continuing processes of housing improvements, increases in density, and changes in land use and service provision were of importance for the design of an upgrading project in George and of interest in the, by then, quite lively international debate on upgrading.

In 1974, the Zambian government and the World Bank agreed on a large upgrading scheme which affected George as well as most other large squatter areas in Lusaka.[3] The third field study, in 1977, was carried out just before the implementation of upgrading. The qualitative aspects of housing remained the prime objective. By documenting eight years of development before the upgrading project, a solid base for a 'before and after' type of evaluation of the upgrading project could be provided. It was also felt that a documentary study of poor people's living conditions was of value in itself (Schlyter and Schlyter, 1980).

By 1980, water standpipes and roads had been provided in George and houses were registered to give the owners security of tenure, but the land remained unsurveyed and in the possession of the city council. Some houses had been affected by the provision of infrastructure, and the inhabitants had been moved to a so-called overspill area (Martin, 1983). Therefore, a block in this area was included in the fourth field study of George (Schlyter 1981, 1984).

The aim in 1980 was to document what the project had meant in terms of changes in the living conditions of the families within the groups of houses studied.[4] In spite of the fact that the upgrading was a physical project, whose main official goal was to improve the conditions of life for the inhabitants, the qualitative impact of the physical environment in the existing settlements was given little consideration in the evaluation studies conducted by the World Bank (Bamberger *et al.*, 1982).

In 1985, the systematic collection of longitudinal data continued, this time combined with a special study of housing conditions among women who were heading households. The need for such a study had occurred when statistics revealed that women householders were under-represented among those who were resettled to the overspill area (Schlyter, 1988; Zambia, 1985).

The most recent field study, in 1989, revealed that many households have improved their houses. However, infrastructure and the general conditions in the settlement had deteriorated and, in spite of hard work and a strong commitment to their houses, house owners were unable to improve their general living environment, and the increasing number of tenants saw no improvements likely in the future. A detailed account of the development during the last twenty years is presented (Schlyter, 1990). By presenting them in full, the data are made available for other researchers as well as for planners and decision-makers in Zambia. Such longitudinal data can be useful in discussions beyond physical planning measures; for example, on adjustment policies and uneven international exchange. Too many decisions are taken on the basis of unfounded assumptions about poor people's living conditions.

METHODS OF INVESTIGATION

The George project is built up by a neat series of six field studies in twenty years, however the study in 1980 broke the four year interval by being conducted ten months too early. There are also other factors that tarnish the image of a perfect study. Because of limited funds, some of the field studies have been carried out during periods as short as two weeks. Consequently, the researchers were forced to concentrate on some aspects of the longitudinal development. For example, in 1980 and 1985, neither the interior of the houses nor the furniture was investigated. The point to be made, however, is that the presence of some gaps in the material is not a severe blow to its ability to provide an excellent picture of the development of housing qualities and living conditions.

The major part of the field work in George was the detailed investigation of six groups of about ten houses each. The focus on the use of space, including outdoor space and the interaction between neighbours, made it necessary to select groups of houses instead of making a random sample of individual houses. The selection was made in order to represent old and central groups as well as those in newer and peripheral areas. With the continuing growth and ageing, all the studied groups have become relatively older and more central to the settlement as a whole. The study of ageing must be part of the research process in any longitudinal study.

The upgrading of George severely affected two of the house groups studied as they were situated on each side of roads subject to widening. Many of the houses were pulled down and the inhabitants were given plots in the overspill area.

All field studies started with meeting the local leaders, they were most knowledgeable on local problems and made the initial introductions to the occupants of the house groups. Subsequent introductions were not necessary as the researchers became known by the inhabitants and, if there were new ones, these were briefed by their neighbours.

A longitudinal study, involving revisits to the same households, creates a special interview situation. Hostility to researchers may be reduced though not all respondents will necessarily allow access to the interior of their houses. But, with continued contact, inhabitants have tended to be extremely helpful and show their appreciation of small gifts; particularly family photographs taken during the previous field study. The most trying aspect of the process is the persistent curiosity of children especially during photographic documentation.

During the first field study the houses were measured in detail. In the subsequent field studies only extended and rebuilt houses were measured. Otherwise, great reliance was placed on photographs. The use of outdoor space was mapped, and the plans corrected with the help of aerial photographs which are available for Lusaka.

Six basic domestic activities were defined: sleeping, eating, cooking, washing, bathing and storing. They were mapped not only on the basis of observations of the activities actually performed but also on what could be understood from the arrangements of furniture, tools and utensils. The measuring of houses is a time-consuming process but allows observations to be made at the same time. In interviews, the location of activities can be checked through questions; informal observations and discussions allow a wider understanding of the spatial aspects of everyday life.

Basic data about the household was collected with the help of questionnaires and the use of previous interview records and drawings of the house. Simple questions on changes regarding the household or the house often generated answers of little use to the interviewer. Respondents often do not, or cannot, remember what they had done in the last four years. However, by looking into data and drawings recorded during previous field work, minor and even larger changes which had taken place could be identified and followed up with questions on when, why, and how. Explanations of, for example, changes in household composition, gave keys to an understanding of the dynamics of the households in their life-cycle. Changes in the use of houses were discussed with the inhabitants in terms of whether they perceived them to be improvements or not.

The combination of interviews and physical investigation in repeated field studies provided different data from contemporary orthodox surveys on, for example, the frequency of tenants and lodgers. By investigating every room and by asking several members of the household specifically about who slept where, there were found to be more inhabitants in each house than was originally stated.

There is a need for interviews to be carried out with both husband and wife. Incomes may seldom be pooled within the households, so that each person should be asked separately about earnings. Also, men and women in George have different experiences and hold different views. For example, before upgrading, women gave priority to clean water, a clinic, and safe space for children to play, while men put roads, larger plots, and transport on the top of their list of priorities. In addition, interviews with key persons, not only local leaders but also nurses at the clinic, teachers, workshop owners, businessmen, and council field staff provided a complement to the views of the 'ordinary' residents in George.

External sources were also referred to; for example, the 1980 census from which, by special order, figures were produced on the level of enumeration areas covering between 100 and 500 households (Zambia, 1985). This made it possible to compare different parts of George, though many inadequacies in the census were revealed. The surveys carried out by the World Bank evaluators also provided quantitative data which could be compared with the small sample in the George study and thereby indicate to what degree it is representative. These studies, however, have largely ignored the qualitative aspects of space.

Aerial photographs have been a priceless tool during field work, for use as maps when investigating services and use of space in larger areas than the house groups. George has been overflown eleven times between 1960 and 1987. Since 1982, there are records at the deeds registry regarding transfer of ownership of the houses.

To summarise, the main components of investigation in the field, which were repeated each time so that changes during the periods could be analysed, are as follows:

In the house groups (60 houses in 1968, 35 still extant in 1989):

1 measuring of houses, indoors and outdoors;
2 photographic documentation of all façades of houses;
3 inventory and photographic documentation of furniture;
4 inventory of building materials;
5 inventory of outdoor arrangements and land use;
6 observations and mapping of use of space;
7 interviews with the heads of households using a questionnaire;
8 informal interviews and lengthy talks with women and youngsters.

In larger areas around the house groups:

1   inventory of urban pattern by looking at use of land, location of doors, latrines and wells.

In the overspill area (from 1980):

1   measuring of selected houses, indoors and outdoors;
2   photographic documentation of the selected houses and their plots;
3   inventory of outdoor arrangements and plot use;
4   interviews with heads of households using a questionnaire;
5   informal interviews with other residents;
6   inventory of building progress in a whole block (200 houses).

In the whole of George:

1   inventory of land use in larger open areas;
2   inventory of commercial and social services, such as shops, schools, etc.;
3   inventory of workshops, churches and other activities;
4   interviews with key persons.

Outside George:

1   search for information in the form of statistics, aerial photographs, deeds in the registry, and reports;
2   interviews with key persons in central government, local authorities and non-governmental organisations.

Additional methods were used on occasion for particular purposes. For example, the deeds register was used to make a sample of women owners, so that women outside the house groups could also be surveyed for a separate, policy oriented, study (Schlyter, 1988) which benefited directly from the knowledge gained from the time series analysis of George. Thus, a longitudinal study may be used to provide baseline data for additional studies as the information which is repeatedly collected has potential for reanalysis for other purposes. This can be illustrated by the following brief discussion of some planning issues.

URBAN PATTERN AND DENSITY

The George study allows analysis of how urban pattern and density have developed in a settlement which was not planned by the authorities. This presentation will concentrate on the logic behind the pattern which emerged, and the qualities it displays. The time series analysis of density in George is based on a combination of aerial photographs, site observations and interviews. The meaning of density for the inhabitants can never be extracted from aerial photographs or surveys alone: observations and interviews are necessary to complement them. The combination of methods and the long time series provides an insight which is more than a simple manipulation of the facts.

Originally, the density was fairly even; newcomers built about 40 houses per hectare. However, a year after, the density had increased to about 50 houses per hectare through infilling. On average, 20 per cent of the land was occupied by buildings. The individual settler proposed the site for his house but it had to be accepted by neighbours and party leaders; thus, a consensus developed on what was an acceptable density. No one complained about density. An individual settler might build his/her house a larger distance from the neighbours, but he/she could hardly protect the large area from being built upon later. The most significant feature of the original urban pattern of George was the absence of demarcation of plot limits. In front of each house was an area of about 20 square metres which was swept and regarded as a private outdoor area. It was the living room and the kitchen of the household, but no arrangements were made to ensure visual privacy. On the contrary, women arranged their places of work so that they had visual contact within talking or shouting distance from each other. In between there was plenty of open space for children to play.

The locations of the houses, doors, and swept areas were carefully considered when a house was built. Different principles co-existed and overlapped. One followed a pattern of free grouping around a common area which was public in the sense that it was overlooked by many. This seems to be derived from rural settlements but the scale was enlarged and the social context different. The interviews clearly indicate that people did not try to reconstruct their communities of origin but were aiming at what they saw as an urban life-style.

The rural pattern of arranging houses was slowly replaced by others; for example, building in lines along existing paths and roads and, towards the end of the 1960s, building in straight lines. This reflects the urban commitment of the builders and also the increasing influence of the party leaders who nourished a vision of a proper and orderly community (see also Martin, 1974). In spite of houses being built in lines, women continued to arrange the outdoor space so that common open areas were created. This pattern cannot be read from aerial photographs but it is immediately obvious from observations on the ground.

The urban pattern of houses was continuously changing even within existing areas. Traffic and heavy vehicles increased as the number of shops grew, many houses along the roads were rebuilt so that the front door facing the road was blocked up while a new door was opened facing a new swept area on the other side of the house. Other owners put up fences to protect their swept areas from the road.

During the upgrading, wide new roads broke up the compact settlement but, between the roads, the original urban pattern was unaffected. A component of upgrading to reduce densities did not have any significant impact. Land was not surveyed as individual private plots but remained in the ownership of the city council so that the original land tenure system was maintained. Land use was flexible and based on users' rights; conflicts between neighbours continued to be solved by the local party leaders who held the same authority as village headmen in the rural land tenure system.

Over the years the land use between the houses became consolidated; trees and hedges now serve as demarcators that cannot be easily moved so that a pattern of plots around the houses is emerging. Houseowners found it wise to claim as much land as possible for future use. Thus the number of fences within the settlement has increased. The outdoor space is enclosed as private space and the reduction of public space is striking; only narrow footpaths remain. However, there is still a certain amount of flexibility: a house owner might incorporate a part of a footpath in his plot by digging it up for cultivation or by putting up a fence. The public will then make a new path by finding the shortest alternative route, probably over a plot belonging to an absent house owner.

Over the years, the density has not increased in terms of number of houses per hectare. Rather, there has been an opposite tendency; small houses have been dismantled. The remaining houses have grown in size so that density in terms of the built area has remained almost constant. The area feels much more dense, but this is because of increased vegetation and more private space. The larger houses are occupied by more people owing to growth of the owner's households and, more significantly, an increasing number of lodgers and tenant households.

The number of persons per hectare has increased from about 250 in 1968 to about 350 in 1987. Consequently, the inhabitants complain of too high a density. Many of them remember the time when the area seemed to have more open space, and everyone can compare their area with overspill and other planned areas where planning standards are very spacious.

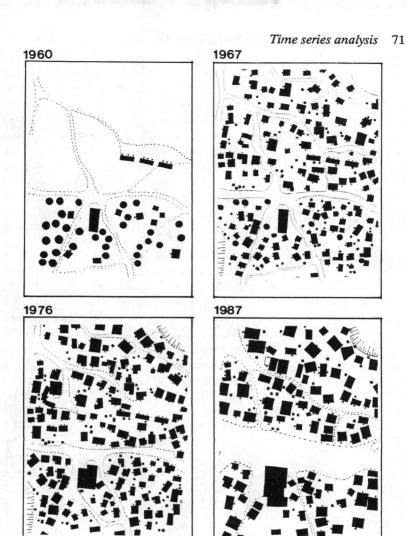

Figure 4.1 The development of a core area in George, 1969–89
These maps are based on aerial photographs and on-site investigations. In 1960
the original farm workers' quarters had been turned into a construction
workers' compound and a number of huts were erected around the large
building which was a grocery in 1969. At its inception, this area was quite
densely built-up with density increasing between 1967 and 1976 owing to house
extensions rather than infilling as in other parts of the settlement. During the
upgrading in 1978, many houses were demolished to make way for the road and
for a 'de-densification' programme. The number of houses has decreased but
the density has increased owing to larger houses with more inhabitants. The
large building has been turned into a bar and much of the land is used for
unloading vehicles. As a residential area, it has become noisy and unpleasant.

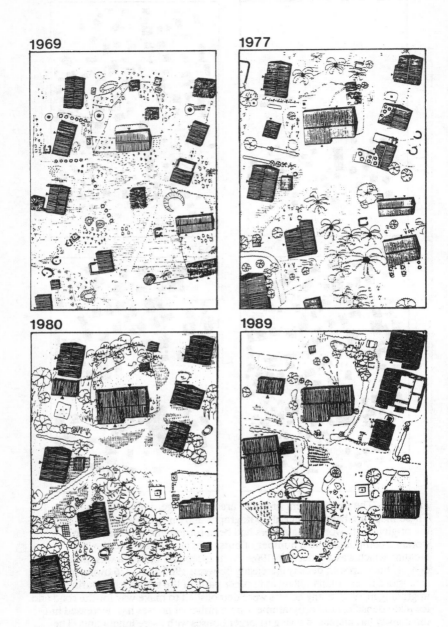

Figure 4.2 Detailed land use maps, George, 1969–89.

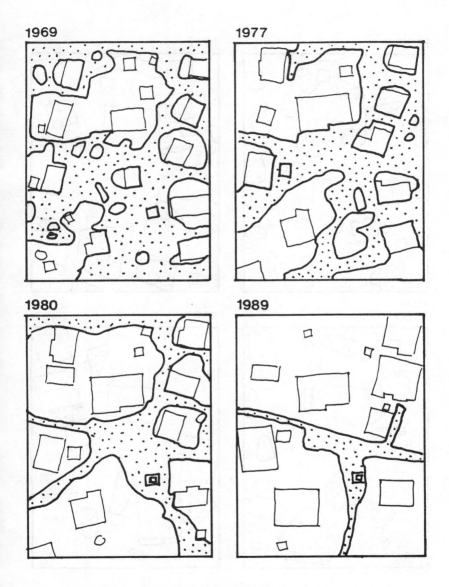

Figure 4.3 Open space accessible to the public and playing children, George, 1969–89. Usable public open space has shrunk over the years from an open area in which houses stood to only narrow footpaths.

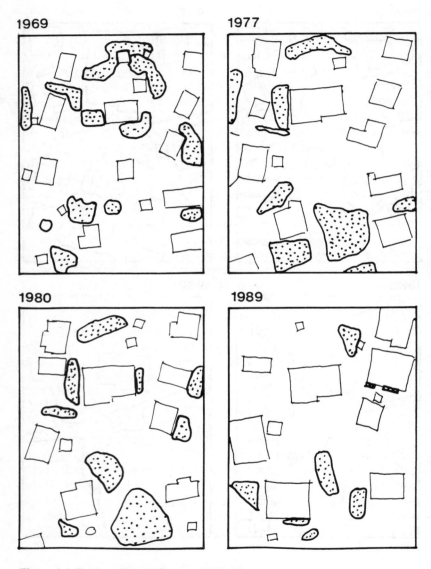

Figure 4.4 Cultivated land, George, 1969–89.
The privatization of space into plots has not increased the area cultivated. Much of the land within plots is eroded and overused by the many residents.

OCCUPANCY RATES AND MODERNISATION OF HOUSING

Occupancy rates, measured in terms of persons per room or roofed area, are often used in evaluating housing conditions and in policy formulation. In discussions of overcrowding, cultural aspects of space use (see Rapoport and Hardie, Chapter 3), the layout of the houses, and the composition of households, have to be taken into consideration. Overcrowding is a normative concept, indicating that the density is higher than it should be according to moral, political, or personal standards. The individual's experience of crowding depends not only on the size but also on the quality of space. The quality experienced is dependent on historical contexts, expectations and ways of life.

A longitudinal study can provide information not only on what people say but also on what they have actually done to their houses over a period. It can also reveal how attitudes alter with changes in the environment, and how statements on building activities are related to actual performance.

The common type of family homestead in rural Zambia consists of a number of one-roomed houses arranged around a swept area. With the urban expansion, the multi-roomed house developed. Newcomers built one or two rooms the first year, and continued to add a room or two during following dry seasons. In spite of small houses and apparently few possibilities of variation, the resultant layouts show considerable variety.

Calculations show that the occupancy rate in George in the 1960s and 1970s was lower than in council rental housing areas, and that it was very evenly distributed between houses and over time. Houses varied in size in George in close correlation with household size. The mean amount of indoor space remained 6.5 square metres per person as houseowners managed to extend their houses with growing families. Only in the 1980s has there been a decline in spatial standard down to 4.7 square metres per person.

Until the mid-1980s, house owners who reported in one survey that they wanted a larger house had usually built their extension before the survey four years later. If overcrowding had been seen as a major problem, they had generally managed to make further extensions. House owners not only extended but also rebuilt and made changes in order to adapt to changing conditions. By blocking up a door or two inside the house and opening up a second or a third entrance door, the owners adapted their houses for rental purposes.

Few people complained of lack of indoor privacy though the change from one family in several one-roomed houses to several families in one multi-roomed house must have created tension. Although life in

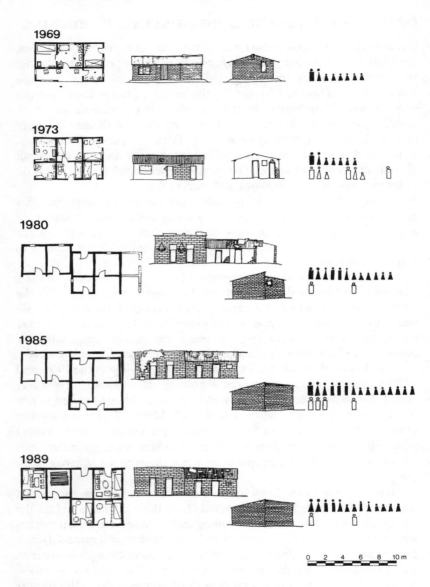

Figure 4.5 A house rebuilt in stages, George, 1969–89.
Built in 1967 and owner-occupied since then, it was extended in the original
mud brick and later rebuilt in cement blocks, part of the old house being
surrounded by new construction and then removed. The number of occupants
has increased from two adults and five children to seven adults (two tenants)
and eight children.

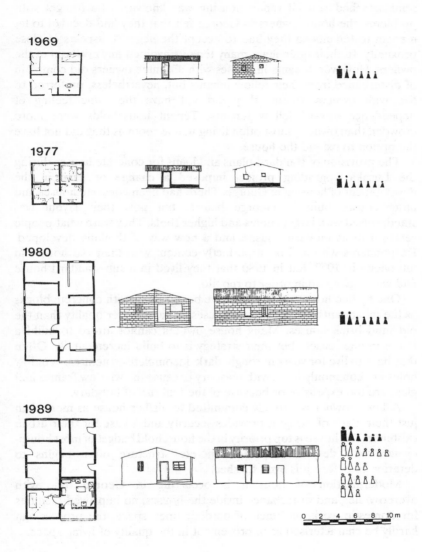

Figure 4.6 Another house rebuilt in stages, George, 1969–89. Built in 1967 and owner-occupied since then, the original mud house with six occupants (two adults and four children) was rebuilt once and for all in cement blocks in 1979. In 1989, it had twenty-five occupants (ten adults and fifteen children) in one owner and four tenant households.

semi-detached council rental housing was known to be fraught with problems, the house owners in George felt that they had decided to let a room to tenants so they had to accept the negative aspects of close proximity. In their interviews, many tenants avoided any criticism of the owners. However, tenants in houses with absentee owners did complain of disturbance from their fellow tenants but, nevertheless, preferred to live with owners absent; they did not have the same feeling of dependence toward fellow tenants. Tenant households were more crowded than owners, most often living in one room as they did not have the option to extend the house.

The provision of standard plans and loans for concrete houses during the Lusaka upgrading project imposed a change of pace in the development of housing in George. There had been concrete houses and large houses built in George before, but now their layout was standardised with larger rooms and higher roofs. They were what people referred to as modern houses, and a new way of thinking developed. Respondents who had been perfectly content with their old houses in interviews in 1977, felt in 1980 that they lived in a sub-standard house and were saving up in order to rebuild.

One by one houses in George have been rebuilt with concrete blocks as the inhabitants consider these houses to be of better quality than the old mud brick houses. Many house owners cannot afford to build a house in one season, but their strategy is to build incrementally. Often they have to live for years in rough, dark, incomplete structures. Window holes are commonly filled with masonry because the window frames and glass are too expensive or because of the real risk of burglary.

A house owner is strongly committed to his/her house as more than just their place of living; it provides security and a base for their urban existence. The house is top priority in the household budget of most house-owners but, despite this, the concrete structure often begins to deteriorate before it is even finished.

Modernisation of housing is occurring in George, but with overcrowding and disturbance inside the houses, no improved facilities for domestic work, and lack of outdoor open space, the process can hardly be characterised as improvement in the quality of living space.

CONCLUSION

The longitudinal method can be used where a constant set of information is collected over several field studies, in combination with any other method, and in studies employing a variety of approaches and theoretical models. The George study illustrates the use of the longitudinal method in physical investigations and analyses of quality of living.

Architects and researchers may discuss quality of space on the basis of their own professional values but, in doing so, it is important to make their values explicit. This can, however, be difficult as values are often well integrated into the entire work. The comparative method is useful for revealing the researcher's own values and avoiding making them the norm (see chapter by Gilbert). The perception of quality of living space, i.e. the values held by the inhabitants about their houses and their settlement, can be compared with the values held by people in other areas. A time series analysis is a special type of comparative study through which, as in the George study, the inhabitants' values can be compared to the values they held four, eight or even 20 years ago.

In all research, and especially in research on subjective matters such as quality, there are elements of interpretation. Conscious of cultural, class, and gender differences, the researcher has to make an interpretation of often contradictory statements from the inhabitants. In a longitudinal study, the researcher can gradually learn more about the community. It is not sufficient to send assistants out in the field; the researcher's own confrontation with the reality of the case study is important in research on quality in order to detect and understand the values of the users. A constant set of information should be collected on each field study but, at the same time, the researcher must maintain an open mind and be prepared to integrate aspects of quality which have not been foreseen and researched in earlier surveys.

The straightforward description of the changes in the physical environment is the first step in the analysis of quality. In common with other longitudinal studies, the core of the George study is descriptive but has, at later stages, been confronted with a variety of questions put within different theoretical frameworks. For example, the questions on commercialisation and gentrification were not on the agenda when the George study was begun. Nevertheless, the descriptive documentation provided a valuable source of information for an analysis of these phenomena (Schlyter, 1987).

Even without more sophisticated analysis, description of changes over time provides better information of people's living conditions than is usually available in planning offices. Innovative planning of housing should build on a knowledge of existing conditions and the cultural heritage which can be provided by simple longitudinal studies which could be carried out relatively cheaply by the planning staff in institutions such as city councils whose concerns and presence are long term.

NOTES

1 This research has been carried out through the good offices of the Zambian authorities and research institutions, in co-operation with the National Swedish Institute for Building Research.
2 In Lusaka, three more townships (suburbs) have been subjected to longitudinal analysis. Muller (1979) made a logitudinal study over nine years in Chawama; Hansen (1982) revisited Mtendere after ten years, and Jules-Rosette (1981) followed the development of Marrapodi over five years.
3 In preparation for the agreement on upgrading, some surveys were carried out within the Zambian authorities, among them one by Carole Rakodi (née White) (Zambia, 1973).
4 The Lusaka Housing Project Evaluation Team produced a series of interesting working papers on the social impacts of the project; for example, Banda (1978), Mulenga (1978), Rakodi (1978), Singini (1978), and Mubanga (1979).

5 Comparative analysis
Studying housing processes in Latin American cities

Alan Gilbert

INTRODUCTION

A great deal of research is carried out to answer a specific problem in a specific context: How many homes have bathrooms? What is the state of repair of the housing? How can that housing be improved? Such research is neither interested in, nor really needs, comparison. Its utility is defined by the answers it produces. It is locally consumed and is useful for its applied contribution.

When too much research of this kind is conducted, however, it could be argued that much of value is missed. In particular, it leads to submission to the blinkers which concentrating on one society is bound to bring. The result is parochialism; something which pervades too much research in most countries. As Walton and Masotti (1976: 2) charge, 'The great majority of urban research has been incredibly parochial and this seems to apply as much to the work of US scholars as to other national traditions.'

The argument of the first part of this chapter is that more comparative research is needed mainly because it helps to reduce the danger of parochialism. Of course, such an argument is only valid if such comparative work is based on sound methods. For this reason the rest of the chapter is concerned with investigating some of the problems that can arise in conducting cross-national and comparative urban research.

WHAT IS COMPARATIVE RESEARCH?

Before trying to demonstrate the value of comparative research, it is necessary to establish what is meant by this term. In this chapter, it refers to work which directly compares two or more cases while employing a very similar methodological approach in each. These cases may be countries, cultures or cities; in this chapter the adjectives cross-national, cross-cultural or comparative are used as synonyms.

Comparative research also includes work which is based on in-depth case studies, or what Ragin (1989: 59) calls empirically intensive research. By this he means research which 'examines many causal and outcome variables in different configurations in a limited number of cases'. It is to be contrasted with what he calls 'extensive, variable-oriented work', research which 'typically examines only a few variables across a large number of cases' (p. 69). Hence, the author is excluding from the term 'comparative research', broad statistical comparisons of data from 60 or a 100 cities or countries. This is not to deny that there is an essential role for such an approach, it is simply not the subject of this chapter.

Within this meaning of comparative research, however, is embraced both research in which the nation or city is the 'object of study' and where it is 'the context of study' (Kohn, 1989a: 20–21). Comparative research can just as legitimately be concerned with studying the cities or nations for themselves as with examining particular processes and phenomena which manifest themselves in those places.

THE VIRTUES OF COMPARISON

The main reason for conducting research in a comparative framework is that it reduces the danger of 'parochialism'. At one level, of course, this is a tautology for one sense of the word parochialism is 'confined to a single, "native" national setting' (Walton and Masotti, 1976: 2).

In a more significant sense, however, comparative work can help to avoid the dangers of parochialism by discouraging the investigator from over-generalising. In particular, it discourages conclusions being made on the basis of experience from one city or country for the purpose of making general statements about all Latin American or even all 'third world' cities or countries. The major advantage, and difficulty, of comparative research is that it regularly reminds the researcher that in certain respects every city is different.

The constant observation that what happens in one city does not automatically happen in another forces the researcher to think harder about underlying processes. Certainly, a comparative approach is not without its own kinds of built-in biases but in this respect it does impose its own intellectual rigour.

Comparing different cities means focusing upon common themes and not being channelled along set routes of enquiry by the specific characteristics of each city. At one level, comparison should concentrate attention upon the general processes. At another, differences between the cities themselves provoke interest: why do such differences exist and what are the implications for the theories and hypotheses under investigation? What is critical about comparison, therefore, is that it forces us to account for the differences and similarities between cities. By itself, the comparative component demands that we avoid mere description. It also cautions us not to make generalisations on the basis of limited evidence, for it brings us into contact with a great variety of housing responses, forms of land tenure, kinds of leader, community organisation, and bureaucratic structure.

In this sense, comparative research involves returning to an earlier tradition of academic enquiry, that followed in at least one major social science discipline – sociology. As Nowak (1989: 34) points out: 'For many classical writers, the notion of "comparative" sociological study would sound redundant. Sociology had to be "comparative", almost by definition.' For Weber, particularly, all research was comparative.

THE DANGERS OF COMPARATIVE RESEARCH

Many studies are called comparative which should not strictly be labelled thus. For example, many books include ten chapters on housing or urban development in ten different cities. Such books have a role but all too often illustrate the dangers either of non-comparison or of superficial comparison. All too frequently, the plea made by the editor to every author to cover the same issues falls on deaf ears. Transportation is discussed in three chapters but given only a few lines in the rest, water supply is examined in excruciating detail in cities where water is scarce but almost ignored in the other chapters. Similarly, such collections tend to lack any real attempt at comparison. The more conscientious editor tries to draw conclusions, noting interesting similarities and contrasts, but the effort rarely reaches beyond the introductory chapter. However good the editor, the basis for a detailed comparison is simply not available. This is not what this author would call comparative work.

Even when in-depth research is carried out using a common methodological framework, there is no guarantee that the results will be valuable. Indeed, in the past many studies built a parochial viewpoint into their methodology; they wore their blinkers into the field, only finding what they were looking for. Indeed, it has often been charged that in the past most researchers from developed countries did just that.

Traditions such as that of urban ecology encouraged investigators to 'venture into "foreign" settings with a prefabricated set of theoretical and methodological tools which presuppose the order and meaning of events' (Walton and Masotti, 1976: 2). Indeed, both Ragin (1989: 60) and Hamnett *et al.* (1984: 97) have gone further in suggesting that such an approach was almost preordained by the empiricist and quantitative tradition of the social sciences. It was certainly common in the attempt to verify such all-embracing theories as 'modernisation' or 'dependency'. 'From the time of Comte, the tendency to array societies along one or a small number of dimensions and to search for one or a small number of basic causal processes has pervaded social science' (Ragin, 1989: 60).

Clearly, this danger is not easy to avoid. One kind of bad researcher will always be imitative and will not adapt methodology and purpose to local realities. Such investigators will find what they seek, even though reality stares them in the face. However, this problem is not confined to comparative research. Indeed, the main case for defending comparative work is that comparison makes it more likely that the investigator will be forced to recognise diversity. More gross generalisations have derived from single case studies than from comparative work.

THE ETHICS OF COMPARATIVE RESEARCH

Cross-national, comparative research is likely to increase the number of investigators working on foreign societies. This is both a virtue and a danger. It is a virtue because it opens the eyes of the researcher to different realities and forms of interpretation. It is a danger because it increases the chances of misinterpretation; in part, because of complexities of language, in part, because of unfamiliarity with the local scene. A particularly delicate issue, however, relates to the ethics of comparative research. Certainly, we can agree with Hamnett *et al.* (1984: 93) that 'ethical problems are more complex in cross-cultural and cross-national research than in research in one's own culture'. These authors are especially concerned with a particular danger of cross-national research; the production of results which are of little interest to the societies being examined. 'It has become evident that cross-national social science activities have been involved in the production of facts and information largely irrelevant to the problems confronting Third World nations' (Hamnett *et al.,* 1984: 115).

While this is a danger, it is not necessarily a greater problem for in-depth comparative research than for overseas research generally. In any case, can we always be certain that foreign investigators are more out of touch with local realities than local investigators? Are researchers drawn from local elites not open to a similar charge?

More relevant is Kohn's (1989b: 93) charge that 'cross-national research has too often been a mechanism by which scholars from affluent countries have employed scholars in less-affluent countries as data-gatherers, to secure information to be processed, analysed, and published elsewhere, with little benefit either in training or in professional recognition for those who collected the data ... the history of cross-national research has not been entirely benign'. Again, however, the problem is probably caused less by the comparative method *per se* than by other factors. In particular, it is the shortage of research funds in less-developed societies, and sometimes of trained researchers, that has increased this tendency. It is much less of a danger today, at least in Latin America and parts of Asia where scholarship is well developed and where funds are now available for local scholars to carry out their own investigations. It is also less of a danger in so far as there is currently much greater pressure on overseas researchers to behave more ethically.

THE BASIS FOR COMPARISON

Some comparisons make sense, whereas others do not. Having made such an obvious statement, it is less easy to state generally when a comparison makes sense and when it does not. The problem is that so much depends on the objectives of the investigation. Since some studies are concerned with social process, others with practical experience, others with underlying trends, etc., each is likely to select different kinds of comparison.

On the whole, however, it normally makes more sense to compare similar rather than totally dissimilar entities or processes. The principal reason for making a comparison is to shed light on similarity of reaction to some common need, on similarities in process, or on how some common result is achieved through different processes. Comparing the housing needs of Brazilian Indians with those of New Yorkers would no doubt make for some fascinating insights into the nature of greed, but it is less than obvious that much will be learned about housing policy for the respective governments. In this respect, we can agree with Abu Lughod (1976: 21) when she argues that there has to be the basis for 'legitimate comparison' between case studies. She recommends

> a strategy which moves in disciplined fashion from the very specific to the somewhat more general to the even more general via the semi-controlled 'experiment', and which attempts to illuminate the similarities and differences uncovered by this research strategy by means of common mechanisms of process.

On this basis, she recommends the approach she employed in comparing three North African Islamic cities: 'looking first at what they have in common, then at the major differences among them, and, finally, at the common processes (applied in variable degrees) which have led to the wide and very real differences we now find' (p. 22).

In the study of housing, this still opens up a range of methodological options. If our main objective is to shame our own government then we will select another country with superior housing conditions as the basis for comparison. If we are concerned with policy, we will select cases where governments have adopted interesting, successful or disastrous approaches. If we are concerned with patterns of consumer demand, we may choose similar kinds of country or society.

As with any research strategy, cross-national research comes at a price: it is costly in time and money; it is difficult to do; it often seems to raise more interpretative problems than it solves. Yet it is potentially invaluable and, in my judgement, grossly underutilized
(Kohn, 1989b: 77)

If the danger on non-comparative research lies in the scholar who sees that everything is unique, the danger in poor comparative research is that nothing strikes the researcher as being different. As Abu-Lughod (1976: 18) argues 'if a theory of urban change and process is to be developed to guide comparative urban studies, that theory will have to steer a more careful course between the Scylla of grand theory and the Charybdis of the pristine case study'.

THE PRACTICAL EXPERIENCE OF COMPARATIVE RESEARCH

During the last 12 years the author has been involved with three comparative studies of housing in Latin American cities. The first study (PIHLU) attempted to compare the ways in which the poor obtained land, housing and services in Bogota (Colombia), Mexico City, and Valencia (Venezuela) (Gilbert and Ward, 1985). The second study compared rental housing and the transition from sharing or renting to ownership in two Mexican cities, Guadalajara and Puebla (Gilbert and Varley, 1990). The third study, which is still under way, is examining landlords, tenants, sharers and government policy in Caracas (Venezuela), Mexico City, and Santiago (Chile). Each study has been collaborative, each has attracted finance from a development agency, and each has employed a questionnaire-based survey involving many hundreds of interviews.[1] On the basis of that experience, some guidelines are provided for those interested in conducting comparative research; at the very least to indicate some of the pot-holes over which the above research has run and into which it has sometimes fallen.

How many cities are to be studied?

In practice, this question is often answered less by the investigator than by the funding agency. In the PIHLU study, for example, the original intention was to compare five cities, so that a wide range of city types could be included. Fortunately, the referees pointed out that this was far too ambitious, given the available resources, and the number of cities was cut to three.

In this respect, Ragin (1989: 57) is undoubtedly correct when he argues that 'it is very difficult to do in-depth research on more than a few cases'; a principal problem being 'that the number of comparisons that must be addressed increases geometrically as the number of cases increases' (p. 59). Certainly, experience suggests that it is not easy for a single team to make a three-way comparison if the aim is to undertake a detailed study of each city. The amount of data to be collected is immense and writing up that material is complicated because there is far too much to be put into a single volume. The PIHLU study produced a comparative volume, a monograph on each of two of the cities, a series of articles, and there is still material to write up.[2]

It is particularly difficult to steer a sane path between the need for generalisation and the need for detail. Enough information has to be provided for the reader to understand the basis of generalisation as well as the nature and reasons for the differences between cities. Clearly, there is an inverse relationship between depth and breadth in a comparative study. Despite the temptation to do otherwise, Kohn's (1989b: 95) declared preference for moderation should be heeded:

> It is not necessarily true that the more nations included in the analysis, the more we learn. There is usually a trade-off between number of countries studied and amount of information obtained ... By and large ... I would opt for fewer countries, more information.

Organisation of research teams

A single researcher can try to achieve familiarity with a range of cases and then attempt to link them in the course of the investigation. This might be termed the Max Weber model of comparative work because in-depth knowledge of many cases is required. An alternative way to accomplish this linking is for a single scholar to take on the role of synthesiser and rely on the expertise of country and area specialists. This might be termed the project model of comparative work because it requires the involvement of a large number of scholars as consultants in a single project. A third way to link case studies is through more intense and more regular interaction among scholars applying similar models or ideas to different cases. This third model, which might be termed the collective model, requires the prior coalescence and ascendance of a relatively small number of guiding questions or research agendas. Isolated scholars working on similar issues applied to different countries could meet to examine commonalities across sets of cases. (Ragin, 1989: 75)

As noted above, the first two models have been used and currently the author is involved in an effort to mount a study using the third kind of approach.[3] Clearly there are advantages and disadvantages involved in each method.

Both PIHLU and the Mexico rental study followed the first model. However, in PIHLU, the team was divided with Peter Ward and one research assistant working in Mexico City, and the author and another assistant working in Bogota and Valencia. In the Mexican study, by contrast, Ann Varley carried out most of the data collection in both cities. The main advantage of using a single team lies in maintaining a common method and approach. If the aim of the study is to explore similar processes, it is essential that the same kinds of methods are used and the same kinds of samples selected. While careful control of methodology is possible with different teams, variations always enter into each team's approach. Such variation is almost bound to be introduced because each team tends to have its own agenda and sets of interests. Without constant liaison and co-ordination, these distinctive interests can easily produce different studies at the end of the day. Even with a split team, problems can emerge for similar reasons, especially where communication between teams is difficult. In the pre-fax days of the PIHLU study, for example, sending a letter from Bogota and receiving the answer from Mexico City could occupy a month, far too long when subtleties of questionnaire and sampling design were being discussed. Certain differences of interpretation were the result.

The main danger inherent in using a single team is that the approach comes to dominate the study and the team may fail to recognise the significance of differences between the study cities. In extreme cases, the data may be collected in an effort to demonstrate the grand vision, rather than the method being adapted to cope with local variations.

Perhaps the principal advantage of separate teams is that local researchers can be involved more easily. Not only do such researchers already know a great deal about their own cities but they often have ready-made contacts in the bureaucracy and in the settlements. Being known in the city is a great advantage, at least in those cities where there are not great schisms based on politics, race, religion, or ideology. For an outsider, particularly a foreigner, getting to know a city takes time. Separate teams are being used in the current study of Caracas, Mexico City and Santiago. Researchers with long experience in the housing field in their own cities are carrying out the field research and doing the bulk of the writing.

The major problem with employing different teams is that it makes co-ordination much more difficult. Even simple tasks sometimes prove beyond solution; once, when travelling from London to a co-ordination meeting in Mexico City the author was told that it had been cancelled while still on the plane!

There are also many difficulties in maintaining a common thread in the research. For a start, the local teams tend to concentrate on what may be a pressing local issue, or on whatever interests them personally, and tend to neglect the comparative element in the research. The co-ordinator on the rental housing project in Caracas, Mexico City and Santiago, is the one most interested in comparison *per se*. His primary objective is to ensure that similar kinds of information are being collected in each city. The individual teams have been interested in the comparison but have been most committed to locally specific issues. There is no necessary contradiction in these different interests but, unless carefully controlled, the local team interests may produce substantially different data sets making comparison impossible.

In the three-city study, we have certainly had the odd sticky patch in establishing an agreed format for collaboration; it might not have been established at all without a number of good meals and the fact that all three team leaders were already personal friends of the author. Differences are now appearing on how the project should be written up. With respect to policy formulation, for example, there was a lively debate in the final meeting about whether a common policy can be devised for cities as distinctive as Caracas, Santiago and Mexico City. Eventually, it was agreed to write both a separate report for each city and a collaborative report on the research as a whole. No doubt the degree of concurrence or difference will become abundantly clear when the comparative study is being written up.

Acquiring similar kinds of information

Any comparative study requires the same kind of data from each city. Clearly, however, the quality of information available is highly variable. In the PIHLU study, good data were far harder to acquire in Valencia than in Bogota or Mexico City. In part, this arose from the fact that Valencia is a provincial city and much of its administration is conducted from Caracas. But, it also stemmed from the fact that the planning office and the local bureaucracy were far less sophisticated than those in the two larger cities. However, there were also problems acquiring good information in Mexico City; particularly because the urban area is administered by two separate governments so that the same kind of material is not always available for both halves of the city.

Similar kinds of problems emerged in the comparison of Guadalajara and Puebla. Guadalajara is divided into three administrative areas and, while most of Puebla falls under the municipal jurisdiction of the municipality of that name, the urban area does spread into several neighbouring municipalities. In that study, it was also hoped to use advertisements in the newspapers to examine trends in land prices over time. Unfortunately, while *El Informador* in Guadalajara constituted a good source of information, equivalent data were lacking in the *Sol de Puebla*. Clearly, comparison is limited to the lowest common denominator.

A further problem arises when different teams are conducting the research, particularly the tendency for each to want to collect different kinds of information to answer locally significant questions. One way of guaranteeing a common source of information for comparative purposes is to carry out a questionnaire survey in each city. This has been a fundamental ingredient in all of the three comparative studies described. Even this method is no guarantee against different forms of data collection. In the three city study, for example, each team wanted to modify the questionnaire and work in different kinds of settlement. In this case, an attempt was made to circumvent the potential dangers by devising a methodology which would collect enough comparable material to guarantee true comparison amongst the three cities while, at the same time, allowing each team to collect information on the locally specific. As such, a method of operation was established whereby each team agreed to apply a similar questionnaire. While they could add any questions they wished, they were not permitted to subtract any. They also agreed to use a similar method for selecting survey settlements, and sampling within them, but they reserved the right, and were given the resources, to pick an additional settlement or two. As a result of this approach, broadly comparable survey data across the three cities has been collected.

Even following a tight methodological approach such as this, the comparative study is vulnerable to other problems. A particular difficulty arises where it is inappropriate or too expensive to make a sample survey of the whole of each city. When surveys can only be conducted in parts of each city, the selection of settlements for inclusion in the survey becomes critical. In the PIHLU study, for example, the aim was to examine how land was acquired, the housing built, and the services provided in low-income settlements. The intention was to examine one recently founded settlement in each city but it was soon realised that much more could be learned by interviewing mainly in a number of older settlements. Such settlements would contain numerous households which had campaigned for and obtained (or sometimes not obtained) a wide range of services. In order to make sensible comparisons between the cities, however, the settlements should be similar. Thus, it was decided to choose settlements with similar levels of servicing and consolidation. Since consolidation is normally linked to age, settlements were chosen in each city with an average age of seven to eight years. A list of service variables was drawn up on which to rank candidate settlements in each city. From this list, settlements with similar rankings could be chosen. Unfortunately, it was soon clear that there were genuine differences in the process of settlement formation and servicing in Bogota and in Mexico City. The major problem was that the size of settlement varied so dramatically in the two cities. Whereas in Bogota a typical settlement would contain between 500 and 1000 households, in Mexico City and particularly in the State of Mexico, the settlements were often vast. This meant that practically every settlement contained a church and a market, features which were comparatively rare in the smaller settlements of Bogota. In the end, different ranking procedures in the two cities were forced on the team.

In the rental housing survey in Guadalajara and Puebla, fewer problems were encountered, probably because they were both provincial cities in the same country. In this study the aim was, first, to learn more about the nature of landlordism and landlord–tenant relations in the two cities, second, to understand the dynamics of residential change among the urban poor, and, third, to examine the transition from tenancy to self-help ownership. To accomplish this it was decided to conduct interviews in three distinctive kinds of settlement. The first kind would be a centrally located rental area containing many large, deteriorated *vecindades*. Many of the tenants in this kind of area would be renting single rooms and sharing communal services. The second kind would be a self-help settlement at least twenty years old and, therefore, well consolidated and serviced; it would also contain broadly similar numbers of owner-occupiers and tenants. The third kind of settlement would be a new peripheral self-help settlement with limited services and few tenants. It could thus be established how new owners had made the residential transition from sharing and tenancy. On the whole, this methodology worked reasonably well. Guadalajara and Puebla are sufficiently similar that similar kinds of urban settlement have developed.

In the three-city rental housing project, however, an attempt to use the same method has proved much more problematic. First of all, urban processes in Caracas, Mexico City and Santiago have proved to be very different. With respect to methods of land acquisition, for example, land has been relatively easy to obtain informally in Caracas and Mexico City but very difficult in Santiago. In the last city, land invasions and illegal subdivisions virtually ceased in 1973 when the government of Augusto Pinochet took over.[4] As a result, there is an absence of new self-help settlements on the edge of Santiago. Those households which manage to become owners, in fact, occupy conventionally built apartments and benefit from 75 per cent subsidies from the government. In practice most of the potential new owners remain as tenants, sharers or 'allegados' in the consolidated self-help settlements. In this respect, a direct comparison between Santiago and the other two cities is simply not possible. In addition, the three cities are much larger than Guadalajara and Puebla. Consequently, it was found that different sectors of each city manifest different residential characteristics. These differences have been exaggerated in both Caracas and Mexico City by the fact that different parts of each city are controlled by different government administrations. This has influenced the form of land alienation and servicing in important ways. As such, the team has been forced to conduct interviews in more settlements than were originally planned. In the end, five settlements were studied in each city.

Design of questionnaires

In theory, the same questionnaire should be used in each city in a comparative study; in practice, certain local modifications must always be made. The degree of modification is likely to increase when different countries are included in the survey, the variation is greatest when different languages are spoken by the subjects of the questionnaire and particularly if the survey is cross-cultural. Within Spanish American cities, there are relatively few problems. Admittedly some words which are thoroughly respectable in one country take on a distinctly coarser meaning in others, so care is always necessary in the phrasing of questions. Similarly, local usage differs so that the appropriate local phrase must be incorporated in each city.

Rather more important is to avoid problems which arise from differences in local sensitivities. All questionnaires touch on certain issues which the interviewees are reluctant to answer. In Colombia and Mexico, few problems were found when asking about political affiliation but, in Venezuela, where politics impinges heavily in most aspects of life, asking the head of household about voting was a sensitive issue. The question was introduced by saying, 'if you don't want to answer the next question, just don't'. Many took up the offer: for while every supporter of the winning party told for whom they had voted, as did the few supporters of the far left, virtually no one claimed to support the main opposition party. In the end, the distribution assumed from this was similar to the official results for the local area; thus, no one was offended but the survey ended up with an apparently accurate answer.

A further complication in Venezuela was caused by the presence of numerous illegal migrants from Colombia. Most lacked official papers and were worried that if they were discovered they would be deported. Hence the totally innocent question about place of birth and former area of residence suddenly assumed hidden meaning. The result was that the proportion of Venezuelans born in the states neighbouring Colombia increased markedly; highland Colombians knew that their accent could give them away so they pretended to have been born in Tachira, a state in the highlands of Venezuela on the Colombian border.

In Mexico, the locally sensitive question related to residential tenure. Landlords did not want to admit that they had tenants, presumably because they were not paying income tax and were afraid that the authorities might find out that they were in receipt of rents. Tenants had also been warned by the landlords to claim that they were friends, relations or guests at the house. In practice, this proved to be less a problem with the questionnaire than with the prior stage of compiling a detailed household count by tenure. The procedure broke down in Mexico as each census indicated virtually no tenants. Only later when the interviews were actually conducted and the individual families were known did the full truth emerge.

Too much should not be made of these difficulties, most questions do not raise hackles. But, unless the research teams are aware that certain questions may be contentious, major damage can be done to the survey.

WHY IS THERE SO LITTLE COMPARATIVE RESEARCH?

In the field of housing or even urban development studies generally, very little comparative research has been carried out in the developing world. Before 1980, there is very little which employed an in-depth comparative approach. The best attempts are represented by the work of Abu-Lughod (1976), Leeds and Leeds (1976), Healey (1974) and Walton (1977). Since then the situation has improved somewhat with comparative work being co-ordinated from the University of Utrecht (Hoenderdos, Van Lindert and Verkoren, 1983), the World Bank

(Mayo, Malpezzi and Gross, 1986), Florida International University (Portes, 1989), and, to a lesser extent, the City Study (Urban Edge, 1988). The emergence of Latin American and developing world networks of researchers has also increased the number of studies following similar approaches and producing broadly comparable kinds of information (Brunstein, 1988; Schteingart, 1988), even if they have not strictly been comparative research projects.

Why is there so little comparative research? The simplest answer is that it is both expensive and difficult to conduct properly. As Kohn (1989b: 77) puts it: 'As with any research strategy, cross-national research comes at a price: it is costly in time and money; it is difficult to do; it often seems to raise more interpretative problems that it solves'. The catalogue of potential pitfalls in this chapter shows the kinds of difficulties which are involved.

Certainly cost is an important factor explaining the lack of overseas comparative research. In Britain, the decline in available research funds has forced most institutions to cut back on larger projects. While they have not been unreceptive to overseas work, admittedly something not believed by most overseas researchers, the high cost of comparative work has put such proposals at something of a disadvantage. Their methods of allocating funds to larger projects have also sometimes been less than helpful to overseas comparative research (Gilbert, 1987a).

But, the main problem seems to lie less with the cost of comparative research than with the tendency of most researchers, except of course anthropologists, to be far more interested in their own society than in any other. Elsewhere, the author has deplored the tendency in British geographical and planning research to equate 'relevance' with the study of the local and the contemporary (Gilbert, 1987a). Few British investigators have ventured beyond the Straits of Dover and, those who have, have tended to exclude themselves from the disciplinary mainstream. In such a context, comparative work has not thrived and, as a result, much of our academic work has shown too many signs of parochialism (Thrift, 1985).

Here, however, is not the place to speculate further on why there has been so little comparative work. It is hoped that it is sufficient to note its absence and to make the plea that there should be more of it.

CONCLUSION

The main value of comparative research is that it is intellectually stimulating. It forces the researcher to eschew simple statements of process by demonstrating that in another city the same facts are explained by different processes or different facts by the same process. Comparison creates confusion, but it is creative confusion. It forces investigators to think harder than they might otherwise have done. Comparison is in no sense a guarantee of good research, but it is a strong stimulus to better research.

Of course, as has been catalogued, comparative research is difficult to carry out. The basis of comparison has to be carefully thought through. Sampling method becomes more complicated when comparison between cities is the goal. Questionnaires have to be modified to match local variations in culture and language. Writing up becomes much more difficult both as a result of the amount of material collected and because of the intellectual problems that comparison poses. But that is precisely the point of comparative research; its difficulties stimulate the mind. It has a built-in tendency to discourage the facile and, if the researcher can avoid the temptation to conclude that all is either wholly similar or is completely different, the full virtue of comparative method is revealed.

NOTES

1 PIHLU was co-directed by Peter Ward, who was then a colleague at UCL, and funded by the Economic and Social Committee for Overseas Research of the British Government's Overseas Development Administration. The Mexico rental study was conducted with Ann Varley, who was appointed to the UCL teaching staff while the study was under way; it was also funded by ESCOR. Finally, the three city rental study has been run in collaboration with local teams led by Oscar Olinto Camacho in Caracas, Rene Coulomb in Mexico City and Andres Necochea in Santiago. It is being funded by the International Development Research Centre of Canada.
2 The books were Gilbert and Ward (1985), Gilbert and Healey (1985) and Ward (1986).
3 We are currently trying to establish a network of investigators interested in rental housing in a number of Latin American cities. This network would encourage each team to carry out their own project but developing similar lines of interest.
4 Only two invasions have occurred in Santiago since 1973. Both were found in 1982 and were supported by the church; they have both been subsequently removed.

6 Analysis of government mortgage records
Insights for state theory and housing policy with reference to Jamaica

Thomas Klak

INTRODUCTION

This chapter describes a rich and relatively untapped source of housing data for developing countries, that of mortgage records at government low income housing agencies. Suggestions are offered on how to obtain access to the confidential data. The advantages and disadvantages of attempting to interpret the data are also delineated. An example is then offered to illustrate the insights into government housing activity obtainable through analysis of mortgage records. In the example, the magnitude and distribution of subsidies among mortgagors of a Jamaican housing agency are quantified, the causes of the subsidies are discussed, and policy changes to increase low income access to housing loans are suggested. The concluding section points to some promising research directions.

A recurrent theme of the chapter is that an analysis of government mortgage records has both theoretical and applied relevance. The research informs state theory by identifying the goals, allegiances, and limitations of the state. Concomitantly, it advises housing policy by identifying mechanisms responsible for low income programmes serving only a limited number of middle income households, and by suggesting modifications.

An analysis of the methodologies employed when using government mortgage records is particularly germane. The choice of internal government computer records as data directs the research along a certain analytical path that contributes much to the outcomes of the research. Government housing records present a classic example of trade-offs between rich insights and frustrating data limitations. This situation calls for particular sensitivity to data issues to obtain the most from a powerful data source, while taking care not to exceed the capacities of the data to inform theory and housing policy.

Concrete examples will demonstrate the insights into state housing policy obtainable through using this technique. The experience of working with data sets from two Jamaican housing agencies, the National Housing Trust (NHT) and Caribbean Housing Finance Corporation (CHFC), is drawn upon for most of the illustrations. It should be stressed, however, that the Jamaican data sets are not idiosyncratic. Many of the same principles of data collection and analysis apply to the housing data of other governments. To illustrate the technique's robustness and some of the international consistencies in the performance of state housing agencies, a few sideways glances will be made to the National Housing Bank (BNH) of Brazil. BNH provided the model for NHT and other governments' housing agencies.

The method should also be relevant to empirical investigations of the performance of other types of state programmes besides housing that are managed via computerised records. These include other types of loan programmes for small businesses, farmers, and recipients of public services. Even more broadly, the experience of analysing government housing records has applicability to government data for other socio-economic variables such as standards of living, consumption, education, and health. For example, the World Bank is engaged in an ongoing international data collection project that includes a great range of household socio-economic variables such as those listed above (STATIN and World Bank, 1988), and the techniques discussed in the paper are likely to be relevant for applications such as the World Bank's data.

GOVERNMENT HOUSING DATA IN THE CONTEXT OF STATE THEORY AND THE HOUSING CRISIS

The source of housing data treated in this paper contrasts markedly with traditional sources. Government computer-stored mortgage accounts replace data from standard questionnaire techniques. The distinction is so great that most analytical opportunities opened by government housing records are unavailable using data obtained through questionnaires.

The research aims associated with the two data sources can be contrasted more precisely. Government data are most useful to research centred on the state (Evans *et al.*, 1985), and can help to define its role as a major player in the housing system. By identifying the actual beneficiaries of the state's low-income housing agencies and comparing them to the general population, the research documents how the state creates institutions that control massive resources and selectively allocate them to certain elements of society. This information speaks to a theory of the state, by situating the state among the classes that it taxes and serves.

In contrast, most questionnaires are administered to households, and, therefore, shed more light on questions such as how the poor house themselves in the informal sector (McLeod, 1987). Most of the poor are in fact disenfranchised from the state, notwithstanding their being in dire need and the target groups of state housing programmes.

When compared, the two data sources answer different kinds of questions about the developing world. Both tell us about housing conditions, but one provides the state's perspective while the other illuminates the poor's viewpoints. They are, therefore, complementary rather than competing approaches to developing world housing. By combining the two methods of data collection, the gap between the needs of the poor and the offerings of the state can be most fully exposed. In terms of policy, a combination of the two methods is most relevant to an understanding of what the poor require and what the state must do to assist them. The degree to which questionnaire and government data do not overlap is not as much a reflection of method-ological differences as it is the state's near exclusion of the poor from its programmes, and thus its databases.

A brief statement of a conceptual framework will provide additional theoretical context for this discussion of methodologies for government housing data (see also Klak, 1990). In most general terms, a state housing agency can be viewed as an input–output mechanism (Offe, 1975). The state responds to demands, attempts to assuage crises, and delivers assistance to the powerful or threatening. Within this general framework, national housing agencies are created where there is a shortage of decent and affordable formal sector housing, and operate by generating mechanisms to accumulate financial resources to be passed on to households to assist with their housing needs.

The research is concerned with the flow of the resources through the state. The resource flow has three components:

1 the *sources* of housing funds, which in many cases are primarily formal employees who have compulsory contributions deducted from their payroll cheques;
2 the *agency's uses of housing resources*; and
3 the *consumption* of housing resources by households, typically through low interest loans.

Each component has its own data sources among government computer files, and these are identified below. The research attempts to trace the funds from the sources, through the state, and to beneficiaries.

A state agency organised in this way controls and selectively allocates massive resources, thereby creating a major system of redistribution, with correspondent political stakes attached. Effectively, some members of society pay for the improved housing (and the associated employment and profits) of others, and the research attempts to quantify this resource transfer.

The example of housing in Jamaica will flesh out the framework presented above. Jamaica's National Housing Trust (NHT) was created in 1976 'to provide a roof over the heads of as many families as possible', with loans distributed according to 'housing needs' rather than by the size of an employee's contribution to the housing fund (Gleaner, 1976a, b; NHT, nd). To finance itself, the NHT has extracted 5 per cent of wages from the formal sector payroll, by pooling contributions from employees and employers (for the same employees). In other words, the NHT extracts about 2.5 weeks of wages per year from each of Jamaica's participating formal sector employees, which currently number about 350,000 in a country with a population of 2.4 million. The NHT's accumulation of contributions has made it Jamaica's third largest financial institution in the country, in terms of assets controlled, behind two international banks, and by far the largest housing financier, public or private. This suggests how large a role state agencies can play in developing world housing systems.

Informal sector workers make up a far larger number than total NHT contributors: for them contributing to the NHT is in law voluntary and in practice rare. The lack of participation by informals owes primarily to the low probability that a contributor will receive an NHT loan: less than 10 per cent have benefited during the NHT's thirteen years of lending. Informal sector workers, amongst whom lies the greatest housing need (McLeod, 1987), are effectively excluded. Although low income *formal* sector employees are included in the payroll deduction programme, they are largely excluded from the loans. For example, despite the NHT's socially progressive goals, an employee in the eighth income decile of the contributor population is more than seven times as likely to have obtained a loan as one in the bottom third of income (NHT, 1988a, b).

Private formal sector housing financiers, of which the most well endowed are Building Societies, Trusts, and Life Insurance Companies, are even more exclusive than the NHT. Housing production and finance is deficient in terms of both quantity and affordability, at an absolute level and relative to the NHT. Output from all sources in Jamaica's formal sector has never even for a single year approached the yearly level that would be necessary for the population to be housed in decent accommodation in the near future (Grose, 1979; Golding, 1982: 6;

'Gleaner', 1984; Jones *et al.*, 1987). Furthermore, a recent comparison of income and housing costs in the private sector has revealed that less than 10 per cent of Jamaican households can afford the least expensive new unit from the private sector (Jones *et al.*, 1987).

Our understanding of the crisis of housing finance would be extended considerably if analyses, comparable to those described in this chapter concerned with government programmes, were applied to private sector mortgages such as those of Building Societies. Owing to their offering more expensive housing and loans with higher interest rates and less manageable terms, and their higher priority on income and savings as loan allocation criteria, private housing financiers serve a much higher income clientele than NHT. Unfortunately, obtaining access to the data of private financiers is even more difficult than it is for a state agency such as NHT (see below).

Clearly, neither the major private nor public financiers accommodate the housing needs of low income Jamaicans. The poor can hope for housing assistance through other government channels, particularly the Ministry of Construction (Housing) and Caribbean Housing Finance Corporation, but these have fewer resources than NHT or the Building Societies.

An overview of the characteristics of government housing data is useful at this point. Greater detail can be found below in a discussion of the advantages and disadvantages of this approach to the study of developing world housing. The immediate concern is with housing agencies' organisation of internal information that is relevant to quantifying the flow of housing resources. Corresponding to the three components of the resource flow through the state described above (sources, state's uses, and beneficiaries), are three types of government data. Of these, the primary data set of interest is one used by the agency to maintain records of housing loan repayment (corresponding to the third component, the beneficiaries). These data contain information about households that have obtained housing assistance from the agency, and about the housing loans themselves. These are the most important data because they describe the resource flow's end point, information which is difficult to obtain from any other source. Thus, most of this paper revolves around these data.

The records that the agency uses to monitor loan repayment contain three types of information:

1 the conditions associated with the origination of the loan (for example, income, loan size, interest rate);
2 the evolution of the mortgage over time (for example, the payment escalation function); and

3 the record of loan repayment (for example, amount of arrears). Within these three broad groupings, the precise variables recorded vary among agencies, depending primarily on what the original organisers of the data set foresaw as important to know during the repayment period.

The other two types of data – for sources and agency use of funds – are important although arguably secondary to the mortgage records. One can obtain a general idea of the *sources of funds* simply by knowing that 5 per cent of the formal sector payroll is deducted per employee, and then consulting published statistics on employment and income. As in the case of Jamaica's NHT, if the contributions to the housing agency are savings for the employee rather than a pure tax, the agency will maintain records of each employee's contributions. These data specify the monetary value of contributions per employee, and because contributions are a direct function of income, can provide a detailed income distribution for contributors.

An assessment of *the internal use of funds* can also be made without access to the agency's full internal accounting records. Pertinent public sources of information include the agency's annual reports and periodic publications, and newspaper reports. Other sources include general information on uses and costs of funds and aggregate investment tables, which can be informally requested from housing agents and experts.

ACCESSING GOVERNMENT HOUSING DATA

A research objective is to obtain a copy of the data sets described in the previous section on tape or cartridge, depending on specifications of the agency's computer system. As in the case of Jamaica, a country may have more than one major government mortgage management agency, each with its own computer system and records. Once copied, these stored data can then be transported to a computer system with full statistical capabilities, as the agency itself is unlikely to have much analytical software. Comparative analysis of the mortgage records of more than one agency, including agencies in different countries, is also facilitated by transferring the data sets to a computer system with statistical capabilities. Additionally, the data analysis will typically continue for many months and so taking a copy of the data away from the agency is highly desirable. The data must be converted to a form that will run on the system with analytical abilities. Conversion of the data storage medium can be tedious.

A state housing agency typically will have many reasons for restricting access to its internal records. Despite what may be implied by the term 'public' agency, its records are anything but immediately accessible to the public. Through restricting information, state agents can protect (a) their clients from breach of confidence and (b) themselves from public criticism. These two defences are obstacles toward which the housing researcher must put considerable and extended effort to overcome. Avenues for overcoming the problem of data confidentiality are discussed next.

Individual data confidentiality is the less difficult problem of the two. The agency can restrict the data set by deleting mortgagor names, identification numbers, and/or addresses so that individuals cannot be identified. This restriction is not a problem as the research is not concerned with individual identification.

Agency fear of public criticism, on the other hand, is a major problem that surfaces throughout the research and, therefore, requires more detailed consideration. Prior to fieldwork, one of the principal reasons to investigate state housing policy is that the agency has not been fulfilling its social mandate of providing low income housing assistance. If the housing system were functioning smoothly, there would be little need for policy revision and little debate over state housing programmes. Instead, the researcher purposefully enters into a social context characterised by unfulfilled promises, dissatisfied constituents, and defensive state agents.

A basic method of self-protection available to the housing agency is to restrict access to its own potentially self-incriminating internal records. The agency itself possesses the most detailed information about its shortcomings. However, the agency may allow a researcher to access the data for the purpose of theoretical inquiry if the analysis also includes output that informs policy. In this way, state theory and state policy overlap. Through discussion with housing agents, followed by a research proposal describing how the data will be used and including the policy-relevant analysis, a set of research guidelines can be identified which satisfy the interests of both parties.

This compromise requires the researcher to consider the investigation in terms of both his/her own goals and the immediate and applied goals of the government agency. This directs the research toward objectives that significantly overlap the two. The inevitably ominous size and complexity of the data set and number of variables make it awkward and inefficient simultaneously to carry out two distinct projects on the same data. Thus caution is advised to ensure that the interests of the agency for descriptive analysis and applications do not pull the investigation too far from the original theoretical framework.

Housing agents can feel more secure about releasing their data to statistical analysis and interpretation if they require review and editing privileges on any writings from the research prior to their being submitted for publication. This may seem to be an overly confining restriction on academic freedom and the researcher's typical independence. However, this agreement may well be the only way to obtain access to the rich government records. Further, state agents in practice seem to have little time or interest in outside academic concerns and are not likely to edit the papers.

Status as a foreigner facilitates access to internal government data files. One is likely to be perceived as less threatening and less politicised, particularly in terms of the possibility of using the data for the purpose of journalistic (as opposed to academic) muck-raking.

ADVANTAGES OF USING STATE MORTGAGE RECORDS

The discussion of the benefits as well as limitations of the data in the next two sections will provide a clearer picture of the fundamental features of the data such as form, size, variables, coverage and potential. Examples of empirical research issues highlight the insights and data limitations.

Large, rich, and unexplored data sets

It would be cost prohibitive, if not logistically impossible, to assemble a data set comparable to those government housing agencies have already assembled on computer file and maintain in an, at least, reasonably current form. Variables which can be either found in the data files or calculated from the data include the date of loan; housing price; loan value; the purpose of the loan (for example, multiple bedroom single family dwelling, core housing unit, home improvements); interest rates; mortgage repayment plans; the monthly payment; subsidies compared to the private market prices and rates; the extent of arrears; mortgagor income, occupation, and gender; and the geography of government housing development. Data of this variety can speak to a great many research questions.

Available data include the entire population of participants in government programmes; no population sampling is required. Furthermore, individual level data, including interval level data for numerical variables, allow for the application of the most precise quantitative analytical techniques. This is a great advance over analysis which has access only to pre-aggregated data (compare Blackwood, 1983). Importantly, access to individual records also allows for exploration of inter-relationships among variables while avoiding the 'ecological fallacy' (where relationships among variables at the group level are wrongfully ascribed to individuals within the groups).

The case of Jamaica will illustrate the size of government data sets. Two government housing agencies, NHT and CHFC, are principally responsible for the collection of mortgage repayments. Between them, their internal records currently contain detailed housing and household data for nearly 40,000 households. This represents the great majority of those that have obtained state housing assistance through loans or grants, and about 8 per cent of all households on the island.

Government mortgage records contain largely unexplored data. Further, because of individual confidentiality, some of the data would be difficult to obtain through questionnaires. Obtaining access to government data circumvents the problem of the individual's unwillingness to disclose personal information. For example, income data gathered through direct survey questioning of individuals is notoriously inaccurate (Personal communication, V. James, Director of the Statistical Institute of Jamaica, July, 1989). Government data for income, on the other hand, are drawn from actual payroll receipts.

Historical record of government housing performance

Through current government mortgage records, one can trace the history of state agency performance in a way that is much more difficult through household surveying techniques or aggregate public data. Housing mortgages typically are repaid over 20 to 30 years. Therefore, although the agency is primarily interested in the current status of its accounts, its lending records contain information that spans up to several decades. Thus one can assess trends in housing costs and loan sizes in relation to household income over time. This provides evidence as to whether state offerings are becoming more affordable or more exclusive over time.

By observing changes in state housing activity over time as it attempts to operationalise its typically ambitious goals and encounters the crisis of the housing system, one gains insights into the relative influence of social groups on the state, the extent of state relative autonomy, and the logic by which the state operates. Typically, state housing loans become increasingly inaccessible over time.

Trends for Brazil's BNH in the secondary cities of Curitiba and Salvador illustrate the cost increases and the insights obtainable through historical analysis of state housing activity. In many years, especially in the 1970s, the real cost of BNH housing was at least four times higher than in 1964–5 when the programme began (BNH, 1985a, b; Reynolds and Carpenter, 1975: 149). Housing cost escalation in part reflected substantial increases in urban land speculation from 1974 to 1980 (Geiger and Davidovich, 1986: 291–2), illustrating the constraints the private market places on BNH. While BNH housing costs were increasing, the purchasing power of the minimum wage was quite steadily decreasing (Maia Gomes, 1986: 280–97; *Brazil Report*, 1981; BNH, 1984: 37; Pastore and Skidmore, 1985: 90–1; *Veja*, 24 June, 1987: 34). Over its history, BNH has become less socially progressive as market and social forces redirected the state's role away from its initial objective of providing popular housing and toward investment in higher income groups (similarly see Pickvance, 1981; Ward, 1986: 55). The populist face of Brazilian housing policy in the 1960s evolved into one that was increasingly market-like in practice as BNH's market responsibilities and Brazil's social context influenced its investments.

Nearness to state housing policy

Research employing government housing records inevitably involves interactions with state agents who set housing policy and manage housing programmes. Deciphering precisely the contents of government data tapes typically requires extended discussions with housing agents who regularly use the data to manage the mortgage portfolio. The researcher is in touch with those powerful in state housing intervention, and the information shared extends far beyond the specifics of the data set to address broader questions such as the difficulties in assisting low income households. The researcher's and housing agents' concerns overlap, even if there are major disagreements over the goals or philosophy of policy or the interpretations of the agency's performance. This interaction significantly informs the research in a way that would be more difficult if data were gathered at the household level.

The analysis of government mortgage records and the interactions between housing researcher and government housing agency may have direct policy implications. Having organised the housing data into a quantitatively analysable form and having become familiar with them, the researcher is valuable to housing decision makers. Consulting opportunities may surface, and the applied work itself informs the theoretical investigation of housing by making additional data available and by providing an insider's view of state housing activity. Adjustments to the delivery systems for housing assistance and improved low income access to housing resources can follow directly from the analysis and resultant recommendations.

Thus there is a strong potential for applied research using government data sources. This is true even if the fundamental research concerns are theoretical debates over the social position of the state, the nature of dependency upon international development agencies, or the limits to self-help housing solutions. One can investigate these theoretical issues while also contributing to low income households' access to housing assistance.

Evidence for state theory

This type of work's potential insights for state theory are numerous. The data offer a rare opportunity to quantify social reallocation of resources by state housing agencies. While studies of the state are often concerned with redistribution, in many cases the most that can be said is that the state has produced an 'income transfer' (Graham and Pollard, 1982). This leaves the question of which wage groups pay and which receive benefits unresolved (for other examples, see Bunker, 1985; Gilbert and Goodman, 1976).

In contrast, quantitative evidence from the government mortgage analysis can be used to assess competing theoretical claims. For example, the state's goals of redistributing housing resources to the poor (a socially progressive goal) and to the most depressed urban areas (a spatially progressive goal) accords with a pluralist conception of the state's role (Lim, 1988). Quantitative evidence of the location and income groups served by state housing investments in Brazil, for example, indicates that the state is much more capable of spatially dispersing resources than it is socially redistributing them. BNH programmes are less socially progressive in the urban periphery where investment has been concentrated (Klak, 1990). This discredits pluralist views of the state as well as others that have collapsed social and spatial issues (Pickvance, 1981; Sayer, 1985; Geiger and Davidovich, 1986). The social distribution of state resources reflects the inequalities of the private market. The evidence of USAID researchers leads them to a similar conclusion, that the public sector cannot effectively address the needs of the poor, as long as a large segment of the middle and upper class cannot find affordable housing (Jones *et al.*, 1987: 85).

DISADVANTAGES OF ANALYSING STATE MORTGAGE RECORDS

This section delineates a list of disadvantages as long as the benefits found in the previous section. However, this should not be taken to imply that the disadvantages loom as large as the benefits. Instead, the extended discussion is a warning of potential problems in the data and aims to note data limitations so that they are not exceeded.

Data errors

One might think that obtaining access to large virtually unexplored data sets for housing and household characteristics would be an unambiguous advantage of the technique described here. Along with the rich potential for uncovering unknown patterns in the housing records, however, comes the difficulty of attempting to analyse raw and often uncorrected data. A significant amount of 'discovery' involves basic data errors to report to the government agency from which the records came, with hopes of their correction for future rounds.

Because the government's housing data have been assembled and computer-entered over many years through the combined efforts of a large number of personnel, there are inevitably erroneous data. These take at least three forms: (1) key punching errors; (2) errors associated with data enterers' misunderstanding of documents or mortgage calculations; and (3) physical deterioration of the data storage media.

For example, the original value of NHT's loans are not the same as the value in the data set as of 1988. NHT's mortgage account manager believed that this was caused in part by physical deterioration of the computer tapes owing to improper storage. Many of the keypunching errors can be discovered by listing the values of a variable of which a realistic range is known. Those cases with values beyond the acceptable range should then be discarded entirely if the erroneous value or values affect others. For example, an incorrect value of monthly payment will make impossible the calculation of the debt-service ratio. When the poor data for one variable do not bear upon the variable being examined, it would not be necessary to exclude the case. Data errors often do not appear to be case specific, but rather scattered widely in the data sets owing to a host of misrecordings.

No data for key variables

It is inevitable that there will be a gap between the agency's purpose of data collection and one's own interests as a housing researcher. The precise variables recorded will also vary among agencies, depending primarily on what those who originally organised the data set foresaw as important during the repayment period. Years later, it may prove impossible to identify the precise meaning or purpose of many variables.

Once the agency decides which variables will comprise the data set, there may be very few if any additions over the years. This is because adding new variables later would require going back to the entries for mortgagors whose loans were made earlier. Because the number of accounts on record for an agency in even a small country like Jamaica is in the 15–25,000 range, there is resistance to adding new information to old records. A result is that data are not collected for all the variables that years of experience in mortgage account management would dictate. The fact that the researcher may be hired as a consultant to the housing agencies' whose internal data files are being analysed for the expressed purpose of informing housing policy is usually not enough incentive to input new variables.

When data are unavailable or inaccurate for key variables such as current mortgagor income or original loan value, surrogate measures must be devised. For example, income from an earlier year can be extrapolated to the present, and current mortgage balance can serve as a proxy for original loan value. Considering negative amortisation of many loans, current balance provides a reasonable approximation for recent loans.

Social distance from housing problems

It may seem a rather straightforward issue that the researcher gives up direct contact with the occupants when government data replace those from household questionnaires. In fact, this social distance is a nagging problem throughout the analysis, and perhaps hinders the housing agency itself as much as the housing researcher. An impression of social insulation is conveyed by state housing agencies, often enclosed in expensive air-conditioned offices and publicly accused of having an 'edifice complex'.

A further distancing from real housing and real people derives from using large data sets. Because there are often many thousands of records, the tendency is to work with the data at an aggregated level, but this can be at the expense of sensitivity to conditions the household faces. In Jamaica, for example, mortgagors might seem comfortable paying an average of only 7 per cent of income toward housing payments. The reality for the household, however, is just the opposite, as massive increases in other consumer necessities, especially food, have squeezed out housing payments and indirectly made shelter unaffordable.

The effect of inflation on the data

A principal aim of housing research using government data is to study the historical performance of housing agencies (see above). More specific issues include ways in which state housing provision moves away from the original goals of the agency, how the regime in power redirects the agency to conform more closely with its philosophy, and how the private housing market influences the agency which often has goals of socially progressive redistribution of housing resources.

Contemporary historical analysis along the lines just described is confounded by the typically high inflation rates in developing countries over the last decade and a half. This creates difficulties for the interpretation of financial data. Inflation itself has had a differential impact on the variables relevant to the research, such as income, housing costs, housing loan value, and the prices of housing inputs and other consumer goods. Adjusting for inflation's impact on these variables is difficult.

Even where monetary corrections for inflation has been a crucial element of state policy, analytical problems abound. In Brazil, for example, where inflation has been so consistently high that the state has instituted inflation corrections on wages, consumer prices, and monthly payments on housing loans from the state, problems are far from resolved. First, several monetary units have been used, and these are not directly comparable. Second, different inflation corrections have been applied to the various items, such as wages and mortgage payments. In fact, BNH's attempt in the early 1980s to increase mortgage payments faster than wages, even though the reverse was true during the 1960s and 1970s, produced a large public outcry against the agency (Siqueira, 1985), and contributed to its elimination (*O Liberal*, 3 March, 1986: 3).

AN EXAMPLE OF ANALYSIS: MORTGAGE SUBSIDIES

This section briefly reviews an example of the kind of investigation made possible by government mortgage records. The subjects of interest are the monetary value of subsidies enjoyed by government mortgagors compared to housing costs in the private market, and the income levels of those receiving subsidies. The case investigated is Jamaica's National Housing Trust. The analysis allows for an assessment of the extent to which the Jamaican government's housing policy objectives, which prioritise low income assistance, are achieved through its mortgage lending patterns.

There are two types of subsidy. *Interest subsidy* is defined as the yearly value of the interest foregone from charging below-market interest rates. The recent Building Society mortgage lending rate of 16 per cent is used to measure the private sector interest rate. *Arrears subsidy* is the yearly interest lost from mortgage payments in arrears (with 16 per cent interest again used as the standard), plus the additional arrears outstanding at the end of the year compared to one year previous.

Mortgage record analysis has revealed that there are J$61 million in total annual interest subsidies at NHT, or J$3,183 for the median mortgagor (about J$5 = US$1 as of May 1988 when the data were recorded). Considering that the median household income in Jamaica is around J$18,000, NHT's interest subsidies represent a major resource transfer.

NHT's interest subsidies are not distributed among mortgagors according to the principle of cross-subsidisation by income level. In other words, higher income households do not pay closer to the value of the housing on the open market so that lower income households can pay less. Instead, the vast majority of the interest subsidies are scattered randomly across income groups. There are large differences in the level of subsidy for mortgagors at the same income level. This means that some mortgagors of a given income level effectively subsidise others at that same income level. Thus, much redistribution of subsidies within the system occurs, but the resource redistribution is not targeted to low income households, and does not increase housing affordability for them.

Contrary to popular opinion, arrears subsidies at NHT are much smaller than interest subsidies. They total J$9.4 million annually, or J$292 for the median mortgagor. The entire amount of arrears subsidies is scattered throughout the mortgagor population, regardless of income level. Also contrary to what might be expected, low income mortgagors do not receive, in absolute value, larger annual arrears subsidies than those of higher income. Large arrears subsidies can be found among low income as well as high income mortgagors; their distribution, therefore, seems to be less a function of income than of the mortgagor's attitude toward repayment to NHT and of housing scheme location (for example, in politically volatile areas).

Loans in NHT's portfolio have been used primarily to finance completed multi-bedroom homes, with some loans for home improvement. Loans have generally been granted to households of middle to upper-middle income. Based on extrapolations, the median income of NHT mortgagors is in the J$30–40,000 range, which is far above the median for the Jamaican population. Thus low income households have not received priority access to NHT's housing loans; there are few housing loans offered for which they can qualify. This owes primarily to their inadequate income to qualify for the housing and mortgages of the size, and with the interest rates and repayment plans, offered.

A number of policy recommendations follow directly from this analysis, of which three can be mentioned here for illustration. First, greater emphasis should be placed on allocating loans for *shelter assistance ranging from serviced lots to starter homes*. These are relatively inexpensive per unit, and thus consume fewer government resources and can assist more households than completed units. Smaller loans are also more affordable to lower income households.

Second, there should be *cross-subsidisation* of mortgagors by income level through progressive interest rates. Despite a policy of interest rates ranging from 4 per cent to 10 per cent depending on income, virtually all NHT mortgagors pay 8 per cent or 10 per cent interest, so interest subsidies depend more on loan size than on interest rate. Few mortgagors earn at the low income levels required for 4 per cent and 6 per cent loans, but this can change by offering smaller loans.

Third, there should be *income ceilings on loans*. While NHT selects applicants based in part on their earning a minimum income necessary to qualify for the loans, it does not restrict applicants that earn much more than needed to make monthly payments. Without income ceilings, higher income households are selected because of their financial strength. This is a major factor explaining why interest subsidies are not progressively distributed. Income ceilings are necessary to allocate loans to the lowest income group that can qualify, and to concentrate interest subsidies there.

SUGGESTIONS FOR RESEARCH DIRECTIONS

The data and research methods described in this paper hold much promise both for expanding our understanding of the role of the state in the developing world and for informing low income housing policy. However, these insights are not easily achieved. A field researcher may spend months interacting with housing agents over the possibility of conducting a quantitative analysis of the agency's housing records. Once the computer records, coding sheets, and hand-written notes are assembled, the more difficult task of interpreting the data still remains. Interpretation requires painstaking record keeping and organisation to keep data sets of tens of thousands of records and dozens of variables in order. Despite these cautions, however, the benefits from analysing government housing records certainly outweigh the difficulties.

The data and methodology discussed in this paper would be even more valuable if integrated with other types of analysis. Three of these suggest some promising research directions. First, mortgage analysis could be fruitfully coupled with institutional analysis which examines more precisely the uses of funds held by a housing agency (see section 2, pp. 97–101). While a quantitative assessment of mortgage records is quite informative concerning the social distribution of housing benefits, it tells us little about how the state uses resources internally before they are converted to housing loans. NHT, for example, has accumulated resources it holds in short-term deposits worth nearly 50 per cent of its total mortgage portfolio. How is it decided how much will be held in short-term investments and who decides it (the housing agency, the central state, the private sector, the IMF)? The state itself may be the principal beneficiary of these short-term investments through accumulating interest on them. Alternatively, private sector borrowers of the funds may reap the greatest benefits.

Second and related to the preceding point, systematic surveying techniques should be applied to enhance an understanding of the housing agents' perpectives. We need to know more about agents' perceptions of the housing crisis and the constraints they face when attempting to operationalise low income housing policy. The housing agents' years of experience in low income housing is an underutilised resource.

Third, the methods described in this paper should be applied to other countries for cross-national comparisons. The examples from Jamaica and Brazil suggest that countries hold a wealth of experience and insights into low income housing policy from which others could benefit. The mistakes made by one country's housing agencies can be avoided in another through the application of comparative research. International comparisons of state housing activity also hold much potential to inform state theory.

7 Ratio analysis

A study of mortgage borrowers in Tunisia

Jean-François Landeau

INTRODUCTION

Mortgage lending raises two financial issues which are interfaced: an issue of credit-worthiness, and another of affordability. The banker will not lend unless he is assured of the credit-worthiness of the borrower. Symmetrically, the borrower should not borrow unless he can afford the debt service on the loan. Such a simple problem can be solved with simple solutions. Ratio analysis, presented in this chapter, is the appropriate tool.

Ratio analysis in mortgage financing has several advantages; simplicity is the foremost. Simplicity is a feature not emphasised enough in the use of statistics. Ratios have a meaning to both lenders and borrowers as measures of the ability of individuals to pile up debt. It is particularly important in housing finance that borrowing and lending decisions should be based on an indicator which can be understood by both sides because it involves numerous one-time borrowers, usually not literate in finance.

This chapter reviews the definitions of key ratios, describes the data source based on a case study, outlines the limitations of these indicators, and provides recommendations on their use.

DEFINITIONS

The most commonly used ratio is the debt-service over income ratio (DSI). It is defined as the monthly or yearly principal and interest payments due on the loan divided by the borrower's monthly or yearly income, and it is stated as a percentage of income. Despite major drawbacks (see below), the mortgage industry is holding to this imperfect indicator because of its simple application. A loan applicant who is unable to set aside up to x per cent of his income for debt-servicing is taken to be not creditworthy.

A less frequently used ratio is the price over income ratio (PI). It is defined as the house purchase price divided by the borrower's annual income and it is stated in years of income.

Because both ratios have the same denominator and the loan is a subset of the house price, they are related. DSI measures the borrower's ability to repay and, thus, deals with liabilities management. In contrast, PI measures the household's ability to afford the ownership of a house and deals with assets management. The balance between assets and their financing plan is maintained through the debt to equity ratio (DE). It is defined as the amount of outstanding debt divided by the borrower's own funds and it is stated as a multiple of the borrower's own funds. It compares the mandatory future savings and the accumulated savings and indicates whether the financing plan is feasible.

Another way to measure the realism of the financing plan and, hence, of the asset purchase decision is to compute the financing gap left after the housing loan (FGAL). It is defined as the housing price minus the loan divided by the applicant's income and it is stated in months or years of income. It measures the down-payment requirement or need for additional finance once the first lender has granted a first mortgage on the property. This indicator allows a cross-check with the DSI and DE ratios.

DATA SOURCES

The preferred source of data for computing these ratios is the loan application filed by households at banks and other credit agencies. To obtain credit, applicants have to disclose their income, the house price, and a financing plan. The often-mentioned doubt about these data is their bias. It is reasonable to assume that data disclosed in loan applications to banks might be overstated in order to obtain loans. Although sceptical (or simply thorough) bankers will always ask for proof of income or will discount undocumented claims, the doubt lingers. At some lending agencies (windows), the incomes claimed in applications must be supported by evidence such as social security contributions or tax returns.

When it comes to income, no data source can claim to be totally accurate. It is a fact that income reporting for tax purposes tends to be biased toward understating. Similarly, the salary reported to the social security system usually excludes *ad hoc* compensation items in order to lower the payroll tax due by employers. The quality of income data is confused further when applicants are working couples. It is even more complex when one works for the public sector and the other for the private sector, each with a different policy about pay and its reporting.

Whatever their drawbacks, the unique features of such data are that they are used by banks' credit committees to grant loans. If it is good enough for a banker, it should be acceptable as a data base for statistical analysis purpose. If ratio analysis applied to mortgage lending is to be meaningful, it must be derived from the data used by decision-makers. This important criterion must be kept in mind when odd results are obtained, such as DSIs well above 50 per cent, PIs above ten years or FGALs up to 50 per cent. They are not the traditional 'outliers' of statistics, but the real cases of loan applicants who have been considered good credit risks by banks or have large accumulations of capital.

In selecting data sources, one must be cautious of proxies when dealing with housing affordability analysis. Specifically, tenants' incomes and expenditure patterns should not be substituted to house purchasers because renters and owners follow two opposite strategies. For tenants, paying a rent is the price for consuming housing services and they prefer this share of their income to be as low as possible. Whereas for owners, housing payments are viewed as the purchase of the largest asset in their life, a security for their family, and this justifies exceptional saving effort. Relying on tenants' behaviour, as exhibited by ratios, can be a misleading proxy to depict the behaviour of would-be house purchasers.

SAMPLING

The minimum size of a sample for ratio analysis purpose is a compromise between the need for a representative size (as high as possible) and the collection and computation time (as short as possible). There are many techniques for achieving the best selection and some rules of thumb can be suggested. In polls, samples of one thousand are deemed representative of tens of millions of voters provided politico-demographic criteria govern the selection process. Similarly, samples of 100 borrowers are generally sufficient for mapping house purchasers' behaviour when borrowers number in the thousands. At the end it is a matter of judgement, that is what the analyst is comfortable with and what he perceives the audience is requiring in order to accept the findings. Two humbling considerations should be kept in mind, however. First, a sample of 100 beats one of zero; and, second, even a population census is not exhaustive. This author has a ready-made offer: if 100 is not convincing, anybody is welcome to duplicate or defeat the results with a larger sample.

Random selection usually allays most concerns. It is thus the most appropriate procedure unless other analysis priorities prevail, such as low-income borrowers, borrowers over 50 years of age, or dual-income borrowers, for whom sub-samples are justified. One affordable pre-caution, when sample size is held at 100, is to run second random samples to test whether key findings are holding. This was the approach used in the case study used to illustrate this chapter in order to secure a better knowledge of owners borrowing from commercial banks and greater confidence in the data quality. More esoteric T-type statistical tests achieve the same result more elegantly. But whatever works for the analyst to gain confidence in his/her data is acceptable.

THE CASE STUDY: MORTGAGE BORROWERS IN TUNISIA

All the above suggestions were implemented during a ratio analysis of mortgage borrowers in Tunisia. The analysis itself was a component of a study of the formal housing finance system done by the author in 1984 (Landeau, 1985, 1987). The purpose was to assess the efficiency of the then complex housing finance system in order to propose remedies. In particular, the objective was to document the ratio policy actually practised at each institutional window to see whether there were con-sistent relationships between debt-service to income ratio (DSI), price to income ratio (PI), and financing gap after loan ratio (FGAL).

In Tunisia, in the mid-1980s, about 15 institutional windows granted mortgage loans. Only two types qualified as financial intermediaries: one savings-and-loans (CNEL) which was limited by its unique contract-ual saving format, and about ten commercial banks. The other windows were one payroll-tax housing fund and two social security funds, one for the private sector (CNSS), the other for the public sector (CNRPS), which offered *ad hoc* responses to CNEL's shortcomings. All windows were recent, none with more than five years of substantial lending experience, and nothing was known of their clientele and their behaviour.

A sample was identified with each type of loan available: for house acquisition, for construction, or for complementing a CNEL loan. There were nine such lending windows as summarised in Table 7.1. It shows that data are dating back to 1983 (gathered for the 1984 study) and to 1985 (updated in 1987 for a follow-up project). Besides the reminder that the example is for illustration purpose, all findings are still relevant. First, they are normalised in terms of percentage of income, years of income, or percentage of down-payment, and thus avoid the erosion effect of monetary data. Second, they are essentially structural ratios which are stable over time unless measured at times of exceptional conditions in the economy.

Table 7.1 also shows the size of the initial sample, which was all the

Table 7.1 Characteristics of the samples of mortgage borrowers, Tunisia

Institution	Share of total housing loans 1977–81 (%)	Window/ loan purpose	Sample date	Initial sample	Final sample
CNRPS		House acquisition	1983	64[a]	64
		House construction		71[a]	71
}	17.3				
CNSS		House acquisition	1983	60[a]	60
		Loan supplement[b]		100	100
CNEL	32.9	Normal loan	1983	339	100
		Advance loan		837	100
		CNSS cofinanced		(100)	(100)
Commercial Banks	8.0	Acquisition	1983	217	100
CNEL		Normal loan	1985	92[c]	92
FOPROLOS	7.3	Acquisition	1985	233	233
Total				2013	920

Note: [a] Entire population of borrowers for the 12 months ending August 1983.
[b] The sample is the same as for CNEL/CNSS co-financed
[c] Entire population of borrowers between July and September, 1985.

loans granted through a window in 1983 (or 1985), and the final size which was either 100 or the entire population, whichever was the least. An exception was made for the FOPROLOS sample because of its low income focus, which justified taking the whole population for analysis. As mentioned earlier, the 217 recipients of commercial bank loans in 1983 were sampled twice.

The income distribution of mortgage borrowers is displayed in quantitative terms in Table 7.2 and in visual terms in Figure 7.1. The same is shown for house price distribution in Table 7.3 and Figure 7.2, respectively. As expected, borrowers at commercial banks were more affluent than those using FOPROLOS or social security funds, but among the latter, public sector borrowers had a median income about 83 per cent greater than those in the private sector and CNEL borrowers overlapped with half of the banks', thus suggesting a bias towards high income. The expected pecking order was obtained with the house price distributions; 'blue-collars' affiliated at the CNSS and commercial banks' borrowers at the extremes of the spectrum. The price range was lower (48:1 between highest and lowest) than the income range (136:1). The proportion of borrowers aiming for 'luxury' houses (i.e. priced above TD 18,000 in 1983 prices) was found to be disturbingly high with regard to traditional affordability criteria (for example, DSI of no more

Table 7.2 Income profiles of mortgage borrowers, ranked by increasing median and date

	Median (TD)	Mean (TD)	Min. (TD)	Max. (TD)	First Quartile (TD)	Third Quartile (TD)	Skewness	Kurtosis	Sample Size
CNSS/CNEL	112	143	39	513	81	177	1.8	4.6	100
CNSS acquisition	140	166	24	568	94	199	1.7	3.6	60
CNEL advances	171	262	67	1541	130	270	3.2	11.3	100
CNRPS construction	220	293	135	1895	180	331	5.0	32.5	71
CNRPS acquisition	256	294	91	862	182	350	1.7	3.3	64
CNEL normal	276	334	73	1027	204	421	1.4	2.1	100
Commercial banks	576	675	167	3266	418	822	3.0	15.5	100
1985									
FOPROLOS acquisition	151	155	100	297	125	176	1.0	1.4	233
CNEL normal	348	398	131	1471	235	475	2.2	7.0	92

Table 7.3 Acquisition/construction price distribution

	Median (TD)	Mean (TD)	Min. (TD)	Max. (TD)	First Quartile (TD)	Third Quartile (TD)	Skewness	Kurtosis
CNSS/CNEL	5600	7792	4000	15750	5330	13000	0.7	-1.3[a]
CNSS acquisition	5992	10161	4000	42540	5308	13014	2.2	6.5
CNEL advances	8250	10629	1580	35000	4905	15522	1.2	0.9
CNRPS construction	11758	13899	7568	45214	9654	15275	2.5	7.7
CNRPS acquisition	12787	14217	4000	45964	5801	19300	1.3	2.0
CNEL normal	15094	13765	1480	31000	7148	18000	0.1	-0.7
Commercial banks	27741	29871	8905	76233	22231	37161	1.1	2.4
1985								
FOPROLOS acquisition	6690	6808	4700	8439	6690	7300	-1.0	1.5
CNEL normal	15675	16953	6870	38000	13720	20353	0.9	1.5

Note: [a] A negative Kurtosis reflects a flat distribution with short tails.

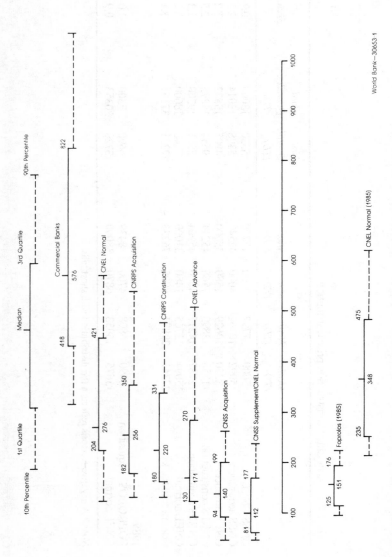

Figure 7.1 Income distribution of mortgage borrowers, Tunisia (monthly income in TD)

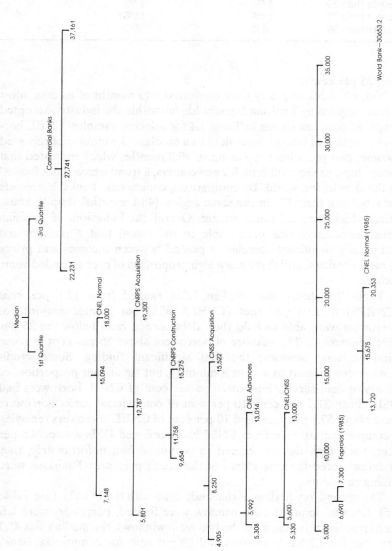

Figure 7.2 House financing windows; price distribution, Tunisia (house prices in TD)

Table 7.4   Price income ratios (PI) as multiples of monthly income, 1983

PI in months of income	Commercial banks	CNEL
Median	48.1	43.9
Mean	55.7	45.6
greater than 49	46%	34%
greater than 59	37%	17%
greater than 79	13%	11%
greater than 99	8%	5%

than 25 per cent).

Yet, when house prices were converted into months of income, most prices targeted by Tunisian households fell within the industry-accepted range. PI data are shown in Table 7.4 for selected samples. CNEL borrowers aimed at houses equivalent to a median 43.9 months of reported income; two years later it was up to 49.8 months, which indicated that ownership was more difficult for newcomers, a trend which is not limited to the developing world. By comparison, commercial banks' borrowers were holding their PIs in the same region (48.1 months) despite house prices which were 2.2 times greater. Overall the behaviour of Tunisian house purchasers was predictable to the extent that a positive and statistically significant correlation existed between incomes and prices at most windows. Still there was a high proportion of over-extended home owners.

Table 7.5 shows that median DSIs ranged from 12.1 per cent (CNRPS) to 51.4 per cent (CNEL/CNSS) but a large majority of borrowers were able to hold their debt-service ratio below the 25 per cent benchmark. The number of borrowers above 35 per cent at some windows was, however, the first significant finding. Some credit committees seemed to accept a minority, but significant proportion, of excessive debt-service burdens. Ten per cent of CNEL borrowers had DSIs above 31.5 per cent; 10 per cent of commercial banks borrowers were above 57.5 per cent; and 10 per cent of CNEL borrowers receiving a complementary loan from CNSS had combined DSIs above 79.5 per cent. Such high figures seemed to be unreal but, unfortunately, they underestimated the true extent of the credit pressures Tunisians were willing to endure.

The second key finding of the study dealt with the FGALs (see Table 7.6). Because loans at each window were limited, borrowers were left with large financing gaps. At the two main windows, the median was 45.7 per cent for CNEL borrowers and 63 per cent for commercial banks

Table 7.5   Housing loans, debt service to income ratios (DSIs)

| | Percentage of income | | | 90th Per-centile | Skewness | Kurtosis | Percentage of borrow-ers below 25 per cent of debt service |
	Median	First	Third				
CNSS/CNEL	51.4	37.9	67.3	79.5	1.0	1.9	5
CNSS acquisition	15.5	9.9	19.6	19.7	−0.6	−0.8	100
CNEL advances	19.1	14.8	27.8	33.5	0.3	−0.5	66
CNRPS construction	20.4	13.3	25.0	26.5	0.1	0.4	53
CNRPS acquisition	12.1	7.2	18.9	22.8	0.3	−0.9	97
CNEL normal	19.5	15.2	27.0	31.5	0.4	−0.5	70
Commercial banks	32.0	22.0	43.6	57.5	0.7	0.6	31
1985							
FOPROLOS acquisition	22.8	19.6	27.3	30.1	−0.1	−0.7	64
CNEL normal	17.3	12.2	24.1	29.0	0.3	−0.8	77

Table 7.6   Financing gaps after housing loans (FGALs), medians and inter-quartile ranges

Medians (25th and 75th Percentile)	Amount (in TD)	Percentage of price	Months of income
CNSS/CNEL	1056	16.3	9.1
	(653, 1888)	(11.6, 19.8)	(5.7, 13.1)
CNSS acquisition	6368	79.7	50.5
	(2975, 13522)	(60.0, 88.3)	(22.1, 83.9)
CNEL advances	302	5.7	3.6
	(270, 4219)	(4.9, 38.3)	(1.7, 17.5)
CNRPS acquisition	10435	75.5	34.6
	(4310, 14703)	(61.8, 80.6)	(19.6, 52.5)
CNEL normal	5683	45.7[a]	19.4
	(2273, 10961)	(37.6, 58.0)	(12.2, 31.4)
Commercial banks	17562	63.0	30.0
	(12080, 22913)	(49.9, 71.8)	(18.2, 43.9)
1985			
FOPROLOS acquisition	1860	27.5	12.2
	(1565, 2012)	(22.0, 30.0)	(9.7, 15.1)
CNEL normal	7486	47.4[b]	22.7
	(5499, 12186)	(40.8, 60.3)	(16.7, 31.4)

Note: [a] After taking into account the borrowers' savings, the median is 10.6 per cent, the first quartile is 5.9 per cent, and the third quartile 38.1 per cent.
[b] After taking into account the borrowers' savings, the median is 17.2 per cent, the first quartile is 5.5 per cent, and the third quartile 36.2 per cent.

Table 7.7　Housing loans: debt service assuming 80 per cent loans (in per cent of monthly income)

	CNSS	CNRPS acquisition	CNEL Normal	Commercial Banks
Median	53.8	35.3	22.8	71.7
1st quartile	31.5	23.3	15.0	47.3
2nd quartile	85.7	53.7	25.6	97.6
Mean	62.3	38.4	21.4	76.9
90th percentile	111.6	66.9	32.5	119.1
Median under Current terms	15.5	12.1	19.5	32.0

borrowers. The other windows were worse on this score and thus indicated their true nature as second mortgage windows. This pattern implied that households borrowing at one major window were also applying for complementary loans at others and getting them. As a result, their combined DSIs were above 50 per cent more often than previously assumed. One piece of evidence for this was the CNEL/CNSS window where households borrowing at two windows had been identified. Their median FGAL was only 16.3 per cent (Table 7.6) but their median DSI was 51.4 per cent (Table 7.5).

Assuming a DE of 4:1 commensurate with the Tunisian saving ability, a test was run on the above samples to estimate the resulting DSIs. The results are shown in Table 7.7 which confirms that most house purchasers had either to produce large amounts of equity, or were combining large debt burdens without the knowledge of the various institutional windows involved. This finding led to recommendations for setting up a housing bank which would finance up to 80 per cent of the house price. The new institution was established in 1989.

LIMITATIONS OF RATIOS

The major attraction of ratio analysis – simplicity – is also its main liability. The drawback of relying on the DSI to accept mortgage credit applications is that it effectively collapses between fifteen and thirty years of repayment risk into one year, the year of income preceding the credit application. Using one observation of a previous year as a proxy for a string of data which can only be forecasted (that is the ability to repay in the future) is intrinsically a weak concept. The DSI as defined above is essentially static with regard to the necessarily long maturities of mortgage loans. However, it will err on the conservative side. If repayment is in the form of constant principal and interest (as is mostly the case except in inflationary environments), the share of income earmarked for debt servicing will decrease over time, thus testing the affordability of the first instalment on last year's income can be construed as the hardest test. The assumption is correct, however, only if the DSIs observed are reasonable to start with, that is in the 30 per cent range. It was shown above that this assumption does not hold in the developing world where households are willing to sacrifice a large share of income to debt service. Whether these efforts are sustainable for the maturity of the loan cannot be answered by a static DSI.

There are several remedies which attempt to introduce the time element in the ratios. The best alternative is to compute a string of DSIs, one for each year of repayment. This could be refined by making the same specific assumptions about income growth, to compute the present value of these income flows discounted at the loan interest rate. The second best solution is to compute another static DSI for each applicant, five years on, thus limiting income projections to a manageable medium-term horizon. If projections are not feasible, a three-year average of past income should be used to strengthen the creditworthiness judgement.

MAKING DECISIONS ON ACCEPTABLE RATIO LEVELS

Ratio analysis is only as good as the numbers attached to it. A PI of five years may be risky or not depending on the level of DSI because the assets may be large, but affordable if purchased with savings. Conversely, a low PI is not a guarantee for the lending institution if the borrower already has substantial debts or financial commitments. The typology of acceptable ratios is illustrated in Table 7.8. Recommended ranges for DSI in the developing world go up to 60 per cent (plus or minus 5 points) when using a moving average of the past three years of income; up to 50 per cent (plus or minus 5 points) with the traditional ratio; and up to 30 per cent (plus or minus 5 points) with the present value of income over the loan life.

Table 7.8 The relationship between price over income ratio (PI) and debt-service over income ratio (DSI)

PI	DSI	
	Low	High (more than 40%)
High (more than 5)	good risk	risky
Low	automatic approval	acceptable but watch for other financial commitments

8 Discriminant analysis
Tenure choice and demand for housing services in Kumasi, Ghana

Kenneth G. Willis and A. Graham Tipple

INTRODUCTION

The classification of 'individuals' into separate groups on the basis of their observed characteristics is undertaken in many fields of housing analysis. The classification may refer to types of spatial area; types of settlement; housing neighbourhoods; house types; housing tenure; household types; housing applicants; or people in housing need.

The term 'individuals' is used here in a statistical sense, to denote any physical object, sociological or economic phenomenon, that is the subject of inquiry. The 'individual' may be a person or a household with certain characteristics; an area with certain features (e.g. a deprived urban housing area).

Classification and clustering analysis is now a commonly used statistical technique to classify individuals into groups on the basis of common characteristics. For example, it has been used to classify geographical areas into types (Webber and Craig, 1978) and residential neighbourhoods (Webber, 1978).

Problems of classification have long been recognised and have been discussed in a statistical sense by Everitt and Dunn (1983) and in a geographical context by Johnson (1976). Essentially they depend upon (a) what criteria are used to classify objects into discrete categories (b) the method of processing the data, for example, statistically minimising sums of squares within groups, or some alternative decision rule; and (c) the level of aggregation in the classification. Varying any of these may vary the discrete classification produced.

This chapter discusses a problem of a fundamentally different kind. Given the existence of two or more groups, i.e. a classification of individual objects into discrete categories; and given that known individuals certainly belong to group 1, and that other individuals are known for certain to come from group 2, what variables best discriminate between these groups, i.e. determine whether an individual belongs to one group rather than another? In addition, how do we set up a rule to allocate further individuals, of whose origins we are uncertain, to the correct group (assuming that they belong to one of the groups in question)? Discriminant analysis is thus distinctly different from clustering analysis: discriminant analysis requires prior knowledge of the classes, usually in the form of a sample from each class. In clustering analysis the data does not include information on class membership: the purpose is to construct a classification.

Although classifications often appear objective and immutable, in reality they are subjective. Events are classified by a variety of criteria, and indeed in everyday discussion some of these distinctions and groupings are sub-consciously, even instinctively made. Everyone's groupings are not necessarily the same, so it is important to get to know what the classification system is before proceeding to try and analyse it.

In doing just that, it is also worthwhile spelling out the purpose of the research discussed in this chapter. Households in Kumasi, Ghana, can be classified into three tenure groups defined below and referred to as owners, family housers and renters. Within each group, housing facilities enjoyed vary between some households with exclusive use of services and those who share facilities such as kitchens, bathrooms and toilets. Service provision is important within a tenure group. Indeed, whether a unit is subject to rent control or not tends to depend upon whether it has shared or exclusive use of facilities. Thus the purpose of this evaluation is to assess what determines whether a household owns, lives in a family house or rents; and within these categories, what determines whether they have exclusive or non-exclusive use of facilities (represented in this analysis by the kitchen). The analysis also seeks to estimate how the distribution of households between these groups would vary if any determining variable (e.g. income) changed.

TENURE AND HOUSING SERVICE CHOICE IN KUMASI

Housing in Kumasi is characterised by privately owned compound houses, usually single storey and about 30 metres square, consisting of rows of rooms ranged around three sides of a rectangular courtyard; with kitchen, bathroom and toilet facilities on the fourth side. Access to all rooms is via the central courtyard which is reached through a single door at the side or rear of the building. Two and three-storey compounds have balcony access to upper floors reached from a staircase in the courtyard. Detailed descriptions and illustrations of the compound house in Kumasi can be found in Rutter (1971) and Tipple (1987), and examples of social life in compound houses can be found in Schildkrout (1978).

Although rental housing appears to have given a reasonable rate of return on investment up to the early 1970s (Schildkrout, 1978), the last 20 years have seen rents fall far behind other prices (Tipple, 1988). The most recent, Provisional National Defence Council (PNDC) Law 138 (and LI 1318), sets the rents for this type of accommodation at 300 cedis per month (2.5 cedis per square foot) per room for those built of sandcrete (cement blocks), and 200 cedis (1.7 cedis per square foot) for 'swish' (houses built of mud). A suite of two such rooms (known as a 'hall and chamber') rent for 400 cedis (1.7 cedis per square foot) or 300 cedis (1.3 cedis per square foot) respectively.

Tenure in Kumasi comprises two basic types: owning and renting. Owning is further divided into two categories: owners and family housers.

About 10 per cent of households in Kumasi own the house in which they live. About half of these have exclusive use of a kitchen and half do not. In Ghanaian culture, ownership of houses, like that of most property, often involves members of a lineage rather than individuals especially in financing such a large structure as a compound. Thus, outright ownership by one individual or household is likely to be clouded by lineage claims. In this way, there are many people who have the right to demand accommodation in a house. Although new laws encourage inheritance through nuclear families and more individual control over self-acquired property (Tipple, 1986), many houses in Kumasi have already been inherited in common by a lineage, under the care of the senior male (*abusua panyin*). Lineage members have a strong claim to live in these houses rent free. Family claims to accommodation are so deeply ingrained in Ghanaian culture that the Rent Act, 1963, allows eviction of tenants if a room is required for a family member. In this chapter, households who live in a family house rent free are referred to as 'family housers'. In 1986 they constituted about 25 per cent of all households in Kumasi. This is a marked growth since 1980 when only 12.8 per cent of households lived as family housers. About 10 per cent of family housers have exclusive use of a kitchen and 90 per cent do not.

Renters are those households who pay rent for a dwelling with exclusive use of services. The type of accommodation renters occupy in Kumasi can be divided between single dwellings built as bungalows and apartments in buildings of two or more storeys (but no higher than four storeys). They can be further divided between government-built dwellings, which are rented out by the State Housing Corporation at subsidised rates, and privately owned dwellings let to private individuals, employees, or major employers for a rent determined by the Rent

Control Committee or by agreement between the owner and the renter according to the quality of the accommodation (rent control has only intermittently been applied to the most expensive houses). In turn, the renter may pay either the rent determined by the Committee or, when paid to an employer, one based on a fixed percentage (usually 10 or 12.5 per cent) of his salary. If the house is a large, well equipped one, the difference between what a worker pays to his employer and what the employer pays to the landlord may be very great indeed, the latter being the larger amount. However, we have no data on the rents actually paid by employers for the accommodation which they rent on to their employees. As both rents and wages are low, many renters pay surprisingly little for some of the best accommodation available in Kumasi (median 300 cedis, mean 324 cedis per month). About 7 per cent of households are renters.

Roomers are classified as those households who occupy rooms singly, or occasionally in pairs (known as a hall and chamber), usually in a compound house, sharing facilities with other households. About 55 per cent of all households in Kumasi could be classified as roomers in 1986.

DETERMINANTS OF TENURE CHOICE AND DEMAND FOR HOUSING SERVICES

Demand for a particular tenure category, with a specific housing service level (e.g. exclusive or non-exclusive use of facilities) might be hypothesised as having the following linear form:

$$q_{thi} = a_0 + a_1 p_i + a_2 p_i^s + a_3 y_i + a_4 l^i + a_5 d_i$$

where

q_{thi} = demand for tenure (t) and housing service level (h) by household (i)

p = price for tenure (t) and housing service (h) paid by household (i)

p^s = price of substitute tenure and housing service available to (i)

y_i = vector of household economic variables, e.g. income, number of workers in the family.

l_i = vector of legal, political, and ethnic qualifications, privileges, and access to land entitlements.

d_i = vector of demographic variables affecting tenure and housing service choice, e.g. age of head of household, male or female headed household, number of children, length of residence.

Some of these variables, and their possible effect on a household's choice of tenure and housing service level, are now briefly discussed, in terms of *a priori* hypotheses, before going on to statistically estimate discriminant functions.

The existence of rent control in Ghana means that the price per unit of housing varies enormously between tenures, as well as with housing services available in each tenure group. A discounted cash flow model, incorporating controlled rents, tax on rents, maintenance, depreciation, land and construction costs, reveals a large negative net present value as far as the landlord is concerned and a negative internal rate of return (Malpezzi, Tipple and Willis, 1990). Decontrolled rents yield a real rate of return of 5 per cent to 8 per cent depending upon the assumptions used. Households in units with exclusive use of services tend not to be subject to rent control and rents of such units are typically three times more than similar units would rent for if subject to controls. Family housers acquiring rights of residence because of cultural obligations on family members or through inheriting property in common on the death of a (matri)lineage member, can live rent free. The price paid for housing reflects in some way a household's preference for housing *vis à vis* other goods; even if, for owners and family housers, this is in terms of an opportunity cost measure.

Demand for housing has been found to rise with income: the income elasticity of demand for roomers in Kumasi is 0.28, which is low compared to others for developing countries reported by Malpezzi and Mayo (1985). Even though rising income across social groups within a society is associated with a fall in the rent to income ratio; across countries, the proportion of income spent on housing rises with increasing income (Malpezzi and Mayo, 1985). The distribution of household income is often quite different for that of per capita income (Datta and Meerman, 1980), especially for very small and very large households but, in both cases, higher incomes permit greater choice in housing tenure and access to better housing services.

Ability to obtain land cheaply has been identified as important for house ownership. Gilbert (1983) discussed the way joining a land invasion in the 1960s and early 1970s allowed relatively poor households a chance to own while households in similar circumstances now do not have that opportunity. Similar time-related opportunities arise through special schemes which make land cheap (e.g. sites and services, opportunities to squat with later legalisation), but those who are not around at the time may never have a similar chance. When government agencies were building houses in Kumasi (chiefly 1948 to 1970), tenants fortunate enough to secure a house (many of whom were war veterans or party faithful), and then to benefit from its subsidised sale to occupants during the late 1970s, obtained housing and land very cheaply.

The traditional land allocation system also allows some households to obtain land on preferential terms on ethnic grounds. A household head whose home village is part of Kumasi can obtain land, on perpetual lease, as of right and with almost no cost. In addition, Asabere (1981) found that non-Kumasi Akans had to pay less to lease land than had non-Asantes. Thus it is expected that Akans, especially those claiming Kumasi as their 'village', would be able to obtain land more cheaply than non-Akans.

A household which has been resident in the city for many years is likely to have absorbed urban values and increased their desire for housing. Also they are likely to have had more opportunities to filter into the type of housing to which its members aspire through time, knowledge, and a network of contacts unavailable to a relative newcomer. It is also evident that land leasing and building a house were cheaper in real terms in Kumasi in past times than now. Thus, length of stay in the city and age of household head would be expected to reinforce a household's likelihood of owning their house.

In most countries, female-headed households would be expected to be able to afford less housing than male-headed households as they tend to have lower incomes. In Kumasi, traditional patterns of non-coresidential marriage complicate this expectation as many apparent female-headed households have a male spouse who contributes to their upkeep as if he were co-resident. On the other hand, his being in separate accommodation may reduce the household's ability to afford better quality housing. Thus the effect of a female household head on housing consumption is uncertain *a priori*.

DISCRIMINANT ANALYSIS AND TENURE CHOICE

Discriminant analysis begins with a situation in which individuals (households in this case) are categorised into discrete groups. Discriminant analysis tries to find a linear function, of n explanatory variables, that provides the best discrimination between the groups (Willis, 1987). It makes intuitive sense that the discriminant function should be chosen so that the variance between groups is maximised, relative to the variance within groups, and this is precisely what a discriminant function seeks to achieve.

Discriminant analysis begins here with the existence of the six tenure groups identified earlier. The object is to find a model or number of variables which will discriminate between these six groups; i.e. correctly assign each household to its existing tenure group. This is accomplished by constructing a linear discriminant function.

$$Z = a_0 + a_1 x_1 + a_2 x_2 + \ldots a_n x_n$$

where $a_1, \ldots a_n$ are positive or negative coefficients of variables $x_1, \ldots x_n$.

Two methods can be used in the selection of variables to include in the model. First, simply include all possible variables and select those which are significant on the basis of a specified statistical test, say an F value. Second, select those variables which, on *a priori* grounds or as indicated by work elsewhere, are thought to determine tenure choice. The selection procedure used in our study involved a combination of both methods: *a priori* variables, including those indicated in Gilbert's work (Gilbert, 1983 and 1987b), were initially selected according to theoretical expectations of their influence on tenure choice, and then only those variables which were significant at a 5 per cent level were included in the discriminant function.

The actual discriminant function or classification is not reproduced. The classification can be based on either individual within-group covariance matrices or the pooled covariance matrix. A likelihood ratio test of the homogeneity of the within group covariance matrix suggested that this method, rather than the pooled covariance method, should be used for our particular data set. The procedure called 'DISCRIM' in the 'SAS' computer package only prints discriminant functions when the pooled covariance matrix is used. When classes do not have normal multivariate distributions, an alternative SAS procedure to DISCRIM can be used. NEIGHBOR classifies observations using a non-parametric nearest neighbour method. Thus, while it is desirable to have continuous variables with normal distributions, it is not a necessary condition. As Kendall (1975) shows, there are cases where discriminant analysis can be applied to qualitative and non-normal data.

The variables which significantly influenced tenure choice correspond quite well with those from Gilbert (1983 and 1987b) discussed above:

1 income (i.e. consumption as a proxy for income) and income (consumption) per capita, which relate to economic factors;
2 age of head of household, length of residence in Kumasi, and length of residence in the house, which relate to adoption of urban values by being in the city for a long period;
3 year of construction of house, and dummy variables for whether a household had lived in a government-built house in excess of ten years (as a proxy for those likely to have been able to buy a government-built house), and for the home region of the head of household, which relate to access to land ownership; and

4 number of rooms in the house, and a dummy variable for condition of house structure, which are unrelated to Gilbert's list.

These variables discriminated between households in the six tenure groups, and produced the results outlined in Table 8.1. The accuracy of the model can be calculated by summing the diagonal and dividing by the total number of observations in the matrix. Therefore, the discriminant model based on the foregoing variables, classified 65.7 per cent of households correctly in their respective tenure groups. The two tenure groups displaying the largest percentage of households correctly classified were owners (exclusive) (78 per cent) and roomers (82 per cent).

There are some other notable features in Table 8.1. First, the discriminant model tries to allocate some households from their respective tenure groups and classify them into other groups, particularly the roomer group. For example, 60 per cent of owners (non-exclusive) are deemed to be more like roomers. Second, a significant percentage of households from renters and owners (non-exclusive) are assessed as being more like owners (exclusive). Third, owners (non-exclusive), family housers (exclusive) and renters are particularly misclassified with this model. A high proportion (48.8 per cent) of family housers (exclusive) were deemed to be more like the family housers (non-exclusive) category. Fourth, overall, the model classified a slightly higher percentage of the housing market as owners (exclusive) (10.5 per cent) than actually occurs in the Kumasi housing market (7.4 per cent); and a similar situation occurs with roomer households. In all other cases the model underestimates households in tenure categories compared with their respective percentages in the housing market. However, we believe that a 66.5 per cent correct overall classification rate achieved in this discriminant model, represents a good equation of the behaviour of households; outperforming prior probability assignment and intuitive judgement by an 'expert' trying to assess tenure choice.

It should not be expected that the model ought to provide a 100 per cent fit: models are abstractions from reality, drawing out the principal features and determinants of particular phenomena. The housing market is dynamic. Some individuals in the renter category may be on the point of becoming owners: hence their characteristics will be more like those of owners, only they have not yet moved between tenures. For some owners, circumstances may have changed so that their socio-economic characteristics are now more like renters or roomers, although they still continue to own especially for cultural reasons.

Moreover, the decision as to what to optimise also determines goodness of fit of the model. Because there are more roomers, the

Table 8.1 Classification summary of Kumasi household tenures (number and percentage)

Posterior tenure cate-gory	Prior tenure category							
	Owners (excl)	Owners (non)	Family housers (excl)	Family housers (non)	Renters	Roomers	Total	
Owners (excl)	39 78.0	0 0.0	1 2.0	1 2.0	0 0.0	9 18.0	50 100.0	
Owners (non-excl)	8 13.8	12 20.7	0 0.0	1 1.7	2 3.5	35 60.3	58 100.0	
Family Housers (excl)	2 9.1	1 4.6	5 22.7	5 22.7	1 4.6	8 36.4	22 100.0	
Family Housers (non)	5 4.1	3 2.4	3 2.4	60 48.8	2 1.6	50 40.7	123 100.0	
Renters	6 15.4	1 2.6	1 2.6	3 7.7	5 12.8	23 59.0	39 100.0	
Roomers	11 2.9	8 2.1	4 1.0	37 9.6	8 2.1	317 82.3	385 100.0	
Total	71 10.5	25 3.7	14 2.1	107 15.8	18 2.7	442 65.3	677 100.0	
Priors	0.074	0.086	0.033	0.182	0.058	0.569		

Note: Owing to the absence of values in some variables, the proportion of each group in the analysis may not correspond to its presence in the population.

discriminant analysis programme tries to assign as many of these correctly as it possibly can, to achieve a better goodness of fit overall. It does mean that a higher proportion of other tenures are misclassified, than could otherwise be achieved.

The above discussion links in with the choice of what *a priori* probabilities should be specified to undertake the analysis and to assign probabilities of group membership. There are basically two choices: specify *a priori* probabilities to reflect the percentage distribution of households among tenure groups in the Kumasi housing market (the approach adopted for the above analysis), or assume there is an equal probability that a household could be in any tenure group. Specifying equal probabilities, using the same variables as in the original model, reduces the overall fit of the model to 48.4 per cent (i.e. the model correctly assigns 48 per cent of households) but the classification rate within all tenure groups except roomers is now improved. The model now classifies 78 per cent of owners (exclusive) correctly (same as previously); 38 per cent of owners (non-exclusive); 59 per cent of family housers (exclusive); 49 per cent of family housers (non-exclusive) (same as previously); and 49 per cent of renters and 45 per cent of roomers. A comparison of these figures with the diagonals in Table 8.1, reveals the improvement in the classification of smaller groups as less emphasis is placed on correctly assigning households in the largest tenure category: roomers. Where the number of households in the sample is equally divided between tenure groups; or where there is an equal chance that a new household entering the housing market could choose any tenure category, then it would be justifiable to use an equal probability criterion of group membership.

Discriminant analysis programmes are useful in providing a listing, for each observation, of the probability of membership of a respective group; with the individual being assigned to the group with the highest probability. Table 8.2 lists some of these results by household identification numbers for Kumasi. The list is not complete, containing only example cases for illustrative and discussion purposes. Nevertheless, a number of points emerge from Table 8.2.

First, the discriminant function confirms the classification of some households with a high posterior probability (e.g. 0201, 1002, 1103) based upon variables in the function; indicating that these households almost certainly represent or characterise the respective group.

Second, reclassification of some households does take place (e.g. 0101, 0102). Occasionally a household (e.g. 0301) originally classified in one group (6) is reclassified with a high probability (0.945) in another group (1). Such instances are rare.

Table 8.2　Classification of households by discriminant analysis

| House & household ID | PC | PDC | Posterior probability of membership in each tenure | | | | | |
			Owners (excl) 1	(non) 2	Family housers (excl) 3	(non) 4	Renters 5	Roomers 6
0101	2	1 *	0.505	0.418	0.000	0.000	0.077	0.000
0102	6	2 *	0.245	0.670	0.000	0.000	0.084	0.000
0201	1	1	0.996	0.003	0.000	0.000	0.000	0.000
0202	6	6	0.000	0.072	0.000	0.075	0.004	0.849
0301	6	1 *	0.945	0.039	0.000	0.000	0.017	0.000
0302	1	6 *	0.104	0.056	0.290	0.151	0.017	0.382
0401	1	1	0.606	0.347	0.000	0.000	0.048	0.000
0501	5	6 *	0.089	0.023	0.213	0.119	0.236	0.320
0601	3	3	0.003	0.000	0.500	0.024	0.469	0.004
0701	2	2	0.231	0.446	0.000	0.000	0.323	0.000
0801	4	4	0.000	0.000	0.000	0.799	0.000	0.200
0901	2	2	0.000	0.642	0.047	0.283	0.000	0.028
1001	4	6 *	0.000	0.005	0.000	0.009	0.001	0.985
1002	6	6	0.000	0.001	0.000	0.013	0.000	0.985
1101	4	3 *	0.128	0.091	0.368	0.074	0.001	0.337
1102	6	1 *	0.599	0.019	0.121	0.037	0.001	0.224
1103	3	3	0.000	0.000	0.998	0.001	0.000	0.001
1104	3	1 *	0.345	0.067	0.269	0.055	0.000	0.264
1105	4	6 *	0.234	0.008	0.222	0.111	0.000	0.425
1106	4	1 *	0.524	0.005	0.184	0.008	0.231	0.048
1107	3	6 *	0.001	0.055	0.328	0.139	0.012	0.466
1201	1	6 *	0.219	0.306	0.067	0.048	0.000	0.360
1202	6	6	0.007	0.197	0.042	0.130	0.101	0.524
1203	6	4 *	0.009	0.029	0.197	0.573	0.000	0.192
1204	6	6	0.032	0.224	0.012	0.173	0.044	0.515
1205	6	6	0.000	0.068	0.021	0.133	0.059	0.720
1206	6	6	0.000	0.142	0.021	0.152	0.029	0.656
1207	6	6	0.000	0.047	0.051	0.175	0.038	0.689
1208	4	6 *	0.000	0.052	0.052	0.201	0.019	0.676
1209	4	6 *	0.000	0.175	0.129	0.216	0.000	0.479
1210	4	6 *	0.000	0.021	0.047	0.097	0.093	0.743
1211	6	6	0.005	0.119	0.133	0.115	0.094	0.533

Notes :PC = Prior classification, PDC = Posterior or discriminant classification.
* = households whose tenure is changed in the model

Third, more commonly, reclassification to another group invariably involves fuzzy cases: cases at the margin of identification with any one group. There seems to be two types of case here:

1 where the similar posterior probabilities involve the prior

classification plus one or two others so that the change is marginal (e.g. 0501, 1104, 1107), and
2 where the similar posterior probabilities ignore the prior (e.g. 1101). (There is also a middle way between 1 and 2.)

The posterior probability across a number of groups is often similar (e.g. cases 0302; 0501; 1101; 1104), indicating posterior indecision as to which group that particular household belongs.

Fourth, the indistinct and uncertain classification of some households in tenure groups and housing service levels should not be really surprising when it is realised that in many respects, and for many of the variables in the discriminant function, the values of the variables are not really distinctive between the groups. For example, the average income (and standard deviation) for owners of both types, roomers, and renters were (in cedis) C18,400 (C12,100), C12,800 (C6,400), and C17,600 (C9,200) respectively. Thus there is a considerable overlap between household incomes in different tenures over much of the income range.

Fifth, Table 8.2 illustrates the complex nature of the household and housing arrangements in Kumasi. For example, in sample house 10 there were nineteen households (although only two are reported in Table 8.2), fourteen of whom were roomers and five were family housers with non-exclusive use of facilities. The discriminant function reclassified all five family housers as being more like roomers, which they probably were apart from the fact that they had inherited a house in common. In house 12, a typical house with eleven households and the owner living on the premises, the owner was reclassified as being more like a roomer. While most roomers were confirmed as such by the discriminant function, one roomer was reclassified as more like a family houser with non-exclusive use of facilities, and three family housers (non-exclusive) were reclassified as being more like roomers. House 11, with seven households and no one owner, reveals considerable reclassification in Table 8.2, and the uncertain classification of individual households by the similar probability of different group memberships.

PREDICTING CHANGES IN TENURE CHOICE AND HOUSING SERVICES

Discriminant analysis can also be used predictively, to assess the influence that a change in one variable will have on the proportion of households in each tenure category. One of the principal variables influencing tenure choice is income.

Table 8.3 indicates how the distribution of households between

Table 8.3   Classification of households as household income increases

Existing household expenditure	Owners (excl)	Owners (non)	Family housers (excl)	Family housers (non)	Renters	Roomers
	10.31	3.61	1.87	14.73	2.68	66.80
+ 20%	11.78	3.48	2.14	14.19	3.88	64.52
+ 40%	13.52	3.48	2.68	13.52	5.62	61.18
+ 60%	15.80	3.35	2.81	12.85	7.10	58.10
+ 80%	18.88	4.15	3.48	11.38	9.10	53.00
+ 100%	21.15	5.09	4.15	10.04	11.51	48.06

tenure categories can be expected to change if income was to increase (in real terms). If income doubled (+ 100 per cent), the percentage of owners (exclusive) would double from 10.3 to 21.2 per cent. The proportion of roomers would decline from 66.8 to 48.1 per cent; while there would be a large increase in the tenant category, from 2.7 to 11.5 per cent of households. Overall, increasing income results in more households in the owner categories; less in the renter categories; and leaves the family house categories unchanged, except that more households here would have exclusive use of facilities. Increasing income results in a much greater proportion of households with exclusive use of services in all tenure categories (14.86 per cent to 36.81 per cent).

Gilbert (1987b) suggests that renting is often a residual tenure, occupied by those households who fail to achieve ownership. Thus, it is important to determine whether there are large numbers of renters who would be owners if they had opportunity to obtain land cheaply. As discussed above, members of local lineages have access to land in Kumasi as of right. A discriminant analysis which assumed all households were natives of the city could, therefore, model this hypothesis. In the event, it showed little change in tenure pattern.

Through discriminant analysis it is possible to examine whether or not it is only a matter of time before appreciable numbers of households will own their housing. This can be done by interfering with the length of stay in the house, the only variable we had which could represent a minimum length of time which a household has been in its current tenure. Table 8.4 indicates how the distribution of households in each tenure category can be expected to change as length of stay increases. Doubling length of residence is associated with a lower proportion of both renters and roomers, indicating that these households would transfer to tenure groups with at least some ownership rights, especially

Table 8.4 Classification of households as length of stay increasess

Existing household length of stay	Owners (excl)	Owners (non)	Family housers (excl)	Family housers (non)	Renters	Roomers
	10.31	3.61	1.87	14.73	2.68	66.80
+ 20%	9.91	4.02	3.75	15.53	2.41	64.39
+ 40%	9.37	4.28	6.43	15.39	2.14	62.38
+ 60%	9.10	5.09	8.43	15.53	1.34	60.38
+ 80%	8.84	6.83	10.31	14.73	1.34	57.97
+ 100%	8.57	8.03	12.72	14.73	1.34	54.63

in family houses (from a sum of 16.6 to 27.5 per cent). The inevitability of individually owned houses passing to joint heirs through mortality probably explains the reduction in owner (exclusive) while any movement to ownership from tenancy is probably mainly into the non-exclusive category.

Both discriminant analysis and multinomial logit analysis can be used to predict the choices of a new individual (household) with given characteristics, and the probability that an individual will choose a different group (tenure), given a change in one of the characteristics. Because of the close relationship between discriminant analysis and the logit model, and also because the logit regression model has far fewer assumptions than the linear discriminant model (SAS, 1986), logit regression might be preferred over discriminant analysis. Even when all the assumptions of discriminant analysis hold, logistic regression is virtually as efficient as discriminant analysis. A logit regression model, using the same variables as in the discriminant model, was developed to predict tenure classification. Overall, the logit model correctly assigned 60.4 per cent of households, a figure slightly lower than the linear discriminant model; and moreover, the logit model failed completely to re-classify households to their existing tenure categories in some groups, concentrating on correctly assigning owners and roomers. Hence, in this empirical case, discriminant analysis is to be preferred to logit analysis.

OTHER APPLICATIONS OF DISCRIMINANT ANALYSIS IN HOUSING STUDIES

Discriminant analysis has a potentially wide application in housing studies, although surprisingly it has been used very little in practice. This section briefly mentions other areas in which discriminant analysis could be used to analyse housing issues.

In financial assessment there is often a need to distinguish between good and bad investments. Discriminant analysis could be used to assess institutional loans and distinguish between good and bad borrowers, in the sense of those who keep up mortgage payments and those who default. Durand (1941) applied the technique to distinguish between good and bad loans for consumer goods, based upon variables such as the size of the down payment, the price of the good, the monthly income of the purchaser, and the length of the contract period. The possibility of using the technique to identify good and bad tenants has obvious implications for landlords! In a similar way, the technique has been used to analyse company and firm failure (Taffler, 1982; Storey *et al.*, 1987).

In housing it may be important in a policy context to analyse tenure change or failure to change tenures. Doling (1973) investigated the household's decision whether to buy the government house in which they were living, or remain as tenants. Variables instrumental in deciding whether a household remained a government tenant or bought the property were the income of the head of the household, the weekly cost of renting, the weekly cost of buying, and the attitude towards owning one's own home. This study was quite rigorous in that other variables, such as the physical structure of the house, which could influence the decision whether to rent or own, remained constant. One of the main difficulties in any quantitative study of housing is that of choosing a method of measurement which takes account of all the variation inherent in housing as a commodity. This difficulty was overcome in Doling's study since the government house is sold to the same tenant: because sitting tenants have the choice of the same physical structure under different property rights, there is no need to measure this physical structure. Doling used this discriminant model in a policy context to predict the proportion of tenants who would opt for ownership with increases in rent levels relative to the cost of buying.

On a spatial basis, discriminant analysis can be used to assist the urban planner in predicting patterns of neighbourhood change. A discriminant model, based upon 90 low income census tracts within the city of Pittsburgh (Fogerty, 1977), predicted 97 per cent of upgrading income paths and 92 per cent of downgrading paths over the period 1960 to 1970, with variables readily available from the decennial population census. A discriminant model in this sense might be a useful guide to policy makers in identifying areas for housing concern. The analysis is capable of extension, to tackle such problems as types of residential renewal programmes appropriate and the optimal points in time for intervention and use of government housing powers. Discriminant analysis is also a suitable technique to test the viability or optimality of settlements selected for development compared to those not selected to receive public investment; and to determine which settlements should be selected for future development (Willis, 1972). The use of discriminant analysis in assessing the significance between the political complexion of local or state authorities and their per capita expenditure on housing or other spatially policy controlled variables might also be noted (Johnson, 1978).

CONCLUSION

Discriminant analysis proved a useful technique in classifying households into one of six mutually exclusive groups, with a 67 per cent success rate. This is much higher than could be achieved by intuitive methods, given that six categories are involved. The analysis of six tenure groups in Kumasi revealed that there are more differences between the households according to whether they have exclusive use of services than whether they own or rent. As in Latin America, owners do tend to be slightly better off and more generously housed than most tenants, although the extended family system gives them large households and the consequent reduction in per capita income. The group we call renters are remarkably similar to owners and appear to find their tenure a valid alternative to owning even for the highest income groups. Renting as a residual tenure can be found among roomers – who are just a majority of households and include many of the poorest and least advantaged – but there is likelihood that many roomers have, or will one day have, a house of their own in their home village. As landlessness is virtually unknown in Ghana, the culture of poverty theory cannot be applied to roomers without major inaccuracies; indeed the owners of many houses occupied by roomers have only slightly higher incomes than their tenants.

The variables related to wealth, ability to obtain land cheaply, and degree of urbanisation, do help to explain tenure choice. The application of discriminant analysis shows that the variables discussed by Gilbert (1983 and 1987b) are influential in tenure choice but there are also some characteristics of the houses themselves which are significant, especially those which separate the lower end of the 'market' from the upper. The evidence that access to services is increased as income increases illustrates that improvements in housing will probably only be achieved through economic development.

Policy-makers should note the polarisation of owners (exclusive) and roomers. Housing policies in Ghana and many other countries have attempted to turn ordinary low income households into owners of self-contained apartments or bungalows. Evidence presented in this chapter suggests that it would be easier to encourage roomers to become owners of compounds where services are shared and rental income is readily available than to try to entice them into the role of 'owner (exclusive)'.

ACKNOWLEDGEMENTS

The authors gratefully acknowledge the generosity of the Leverhulme Trust in providing a grant under which this work was funded, the World Bank for permission to use the 1986 Kumasi data set, and Alison Prince for statistical computations. All views and any errors are those of the authors.

9 Regression analysis

Determinants of overcrowding and house condition in Ghanaian housing markets

Kenneth G. Willis

INTRODUCTION

Frequency and cross-tabulation tables have limitations in their ability to analyse housing problems. Suppose a sample of houses consisted of 180 observations, and the purpose is to investigate the effect of income, household size and tenure type on housing consumption. By cross-tabulation, the sample could be divided into, say, five income groups, four household size groups and three tenure groups. Means could be computed from each cell to estimate the effect of these variables on demand. However, with $5 \times 4 \times 3 = 60$ cells, there would only be an average of three observations per cell, and although some would have five or six, many would be empty and most would have only one or two observations, making means meaningless or extremely unreliable (Malpezzi, 1984b). Regression techniques get round this problem, in addition to that of errors in results and interpretations which can occur in cross-tabulations through the inappropriate grouping of survey data (Upton, 1989).

Regression analysis can be broadly defined as the analysis of the statistical relationship among variables. There are many forms of regression analysis but a basic distinction can be made between forms which use continuous data and those which use discrete data. The classical regression model is based upon a continuous dependent (response or endogenous) variable (Y) which is dependent upon a number of independent (predictor or exogenous) variables (X_1, X_2, X_3...., X_k). Such a model implies cause and effect; a given change in X_1 will cause Y to change by a specific amount; but it does not prove cause and effect. It really stresses the probabilistic association between the two variables. Indeed instrumentalism argues that variables and models should be chosen which provide the best predictive results (since this is the primary purpose of a theory) rather than selecting those based on deductive logic or causal theory (Boland, 1979).

SIMPLE LINEAR RELATIONSHIPS

A simple one variable model in which the relationship between the dependent variable (Y) (housing expenditure) and the independent variable (X) (income) can be written as a linear equation

$$Y = a + bX$$

Figure 9.1 shows a number of observations between rent and income. The equation which represents the data is

$$Y = 0.25 + 0.5X$$

If a prediction of housing expenditure is required when income = 6, then, if 6 is substituted into the equation in place of X, and the equation solved, the value of Y is 3.25. The slope of the line (b) indicates how responsive Y is to a change in X.

Usually data consist of observations as in Figure 9.2, where there is a relationship between X and Y, but not an exact one. Regression analysis attempts to fit a line through these points, such as to minimise the sum of squares of the vertical deviations from the line to the actual points.

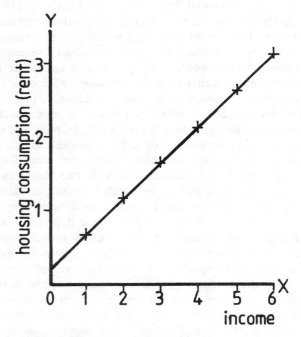

Figure 9.1 A strict deterministic relationship between two variables

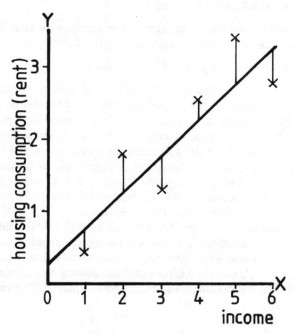

Figure 9.2 A statistical relationship between two variables, minimising sums of squares

The reason for minimising the sum of squares rather than just their sum, is that some differences between the observed values of Y and the predicted (line) values are positive and some negative. If only the sum of the deviations was considered it would be possible for this sum to be zero even though the deviations were quite large. Actual methods of calculating the values of (*a*) and (*b*) which give a least squares line are documented in many standard texts (see for example Draper and Smith, 1981; Wonnacott and Wonnacott, 1985; Rawlings, 1988). For Figure 9.2, the relevant equation is

$$Y = a + bX + u$$

where (*u*) is the residual or difference between the value predicted by the estimated (*a*) and (*b*) for a given observation of X and the actual value of X for that observation.

MULTIPLE REGRESSION

One of the main advantages of regression analysis is that it is a multivariate technique. Multiple regression of the form

$$Y = a + b_1X_1 + b_2X_2 + b_3X_3 + ... b_kX_k + u$$

can be used to sort out the separate effects of several explanatory variables. The advantage of this technique over cross-tabulation can be appreciated by returning to the example in the introduction. With income entered as a continuous variable (X_1); three household size dummies of 0 or 1 (X_2, X_3, X_4); two tenure group dummy variables (X_5, X_6); and a constant (intercept) term (a), a regression to estimate the separate effects of each variable has $180 - (1 + 3 + 2 + 1) = 173$ degrees of freedom. Estimates based on this method will be more reliable than the cross-tabulation method, using the same data.

Computer packages such as SAS, SPSSX, MINITAB, BMDP, etc., have regression procedures and programs which rapidly calculate results. Statistics associated with regression results are parameter estimates $(b_1, b_2,...b_k)$, standard errors of these estimates, t statistics for the hypothesis that the coefficient is zero, and R-squared (R^2) values.

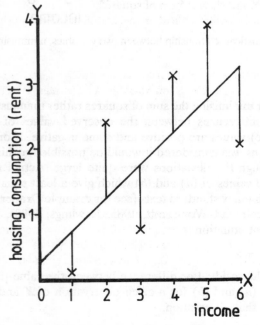

Figure 9.3 Another degree of scatter, influencing R^2 value

Formulae and computational details for these can be found in statistical texts. The purpose here is to provide some intuitive understanding of these measures. A comparison of Figures 9.2 and 9.3 shows that while each line has the same slope and intercept, data points are clustered around the regression line in Figure 9.2; whereas in Figure 9.3, while the line is still the best fit which minimises sums of squares, the data do not fit the line so well. The same regression equation (intercept and coefficient) fits the two data sets, but the R-squared statistics are very different. R^2 provides a measure of the goodness of fit of the regression equation; its value lies between 0 and 1. In Figure 9.1, $R^2 = 1$. There is no error in the data: 100 per cent of the variance in the data is explained. If R^2 is small, is the data unrelated, or is the relationship a true one but one which is weak? The test of whether R^2 is due to chance or not can be assessed in terms of an F value. R^2 is derived from

$$TSS = ESS + RSS$$

where

TSS = total sum of squares of the dependent variable

ESS = residual or error sum of squares

RSS = explained variation Y or regression sum of squares

$$R^2 = 1 - \frac{ESS}{TSS} = \frac{RSS}{TSS}$$

R^2 is the proportion of total variation in variable Y that is explained by the regression of variable Y on variable X_1 (or all Xs in the multiple regression). When $ESS = 0$, $R^2 = 1$ as in Figure 9.1.

If the number of observations is small or the number of independent variables is large, a large R^2 can result simply because there are not enough data or pieces of information, to assess whether parameter estimates are due to chance or whether there is a significant relationship. Drawing a line through two or three data points, will always result in an R^2 close to 1, simply because there are no other data points left over to indicate whether there is any error or not. A test of whether an R^2 is due to chance or not therefore requires the 'degrees of freedom' or number of observations in excess of the number of regression coefficients to be taken into account. The test of significance F is

$$F = \frac{R^2/k}{(1-R^2)/(N-k-1)}$$

where k = number of regression coefficients, N = number of sample observations

An F table in any statistical text (see for example, Siegal, 1956; Barber, 1988) will indicate whether the value is significant and at what level; for example, if there is only a 1 in 20 chance (5% level) that we will not reject the hypothesis that the result is *random*, when it is in fact random.

In many applications of regression analysis, the ultimate aim is to arrive at a good estimate or prediction of Y using variable X. Success can be measured by the degree to which variable Y can be predicted accurately. But each coefficient estimate is a random variable: every time a new sample is drawn from a given population for the independent variable or variables and a new regression run, different coefficients result, even if the underlying population parameter, the true coefficient, remains fixed. However, over a number of samples, the 'average' coefficient would approximate the 'true coefficient'. The standard error of the coefficient, defined as the standard deviation of the residuals about the regression line, provides a measure of the distribution of these errors.

Intuitively, a small standard error relative to the coefficient estimate indicates an efficient and consistent, or good, estimator. The t statistic, or the estimated coefficient divided by the standard error of the coefficient, provides a test of whether the coefficient is really significantly different from zero. Again t test significance tables indicate, on the basis of degrees of freedom and a probability level (for example, 1 chance in 10, or 0.1; 1 chance in 20, or 0.05; 1 chance in 100, or 0.01), if the t statistic is greater than the critical table value, and hence, whether the coefficient is statistically significantly different from zero, that is, it is not likely to be a random chance occurrence.

VIOLATIONS OF ASSUMPTIONS IN THE ORDINARY LEAST SQUARES REGRESSION MODEL

The ordinary least squares (OLS) regression model is based on a number of assumptions about the way in which the observations are generated (Koutsoyiannis, 1977; Kennedy, 1979). These assumptions are:

1 the dependent variable is a linear function of a set of independent variables plus an error term;
2 the expected value or mean of the distribution of the error term is zero;

3 error terms have the same variance and are not correlated with one another;
4 observations of the independent variables can be considered fixed in repeated samples.
5 the number of observations exceeds the number of independent variables and there is no linear relationship between the independent variables.

Violations of these assumptions lead to parameters and results which are biased, inefficient (have a large variance), and inconsistent (vary as sample size increases). However, when violations of these assumptions occur it is frequently possible to correct for them and still derive unbiased, efficient and consistent estimators. Some of the more important violations and corrections for them, are discussed below; although, again, statistical and econometric texts provide a more detailed and comprehensive coverage.

Violations of the first assumption are termed specification errors and involve:

1 wrong regressors: the omission of relevant independent variables and/or the inclusion of irrelevant independent variables;
2 non-linearity: the relationship between the dependent variable (Y) and independent variables $(X_1, X_2...., X_k)$ is not linear,
3 changing parameters: the parameters (b) do not remain constant during the period the data are collected.

Violation of the second assumption, perhaps arising because of an omitted variable, gives rise to a biased intercept. This may not be a problem in some instances, in that the error term is incorporated into the prediction, and the slope coefficients (b) remain unaffected. The two major problems associated with violation of the third assumption are:

1 heteroskedasticity: errors do not all have the same variance;
2 autocorrelated errors: errors are correlated with one another.

Violations of the fourth assumption centre around:

1 errors-in-variables: errors occur in the measurement of the independent variables;
2 autoregression: the use of a lagged value of the dependent variable as an independent variable;
3 simultaneous equation systems: since at least one endogenous variable is an independent variable, it cannot be considered fixed in repeated samples, and any change in an error term will change all endogenous variables.

Violation of the fifth assumption is called multicollinearity, and occurs when two or more of the independent variables are linear functions of one another in the sample data.

The omission of a relevant variable in a regression equation will result in biased estimates of the remaining coefficients in the equation (unless by chance the omitted variable is orthogonal to the included variables) and a biased intercept, unless the mean of the omitted variable is zero. The inclusion of irrelevant variables does not bias coefficients, but may lead to less efficient estimates. Specifying the correct set of independent variables is not easy. Unless they can be justified by theory, variables should not be included in the set of independent variables. An appropriate set of independent variables can then be determined according to statistical inference.

Figure 9.4 shows that, for a non-linear functional relationship, OLS regression parameter estimates will be biased and also meaningless. However, independent variables can be transformed to make them linear. OLS requires the relationship between each coefficient and its variable to be linear. Variables themselves may be non-linear. Assume the demand for rooms is a function of household size. However, as

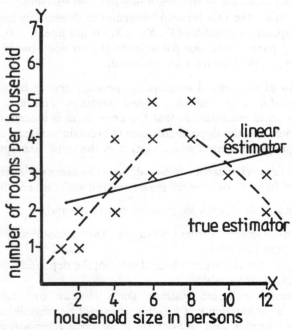

Figure 9.4 A curvilinear relationship

household size increases, the number of rooms demanded does not increase in a linear fashion: a household of four does not usually demand twice as many bathrooms, kitchens, dining rooms, etc., as a household of two. As household size increases, demand for rooms increases at a declining rate, as in Figure 9.4; and may eventually fall, if large households through poverty are forced to live in one or two rooms. A linear estimator would be a poor predictor of the number of rooms desired by a household, and would have a low R^2. However, a curved relationship can be transformed to a linear one by including the square of the variable.

$$Y = a + b_1X_1 + b_2X_1^2 + u$$

Such an equation would improve the goodness of fit, with larger t statistics and a higher R^2 value. Logarithmic, reciprocal, and other transformations are also typically used in transforming data to estimate demand functions (Willis and Benson, 1988).

If the relationship between an exogenous and endogenous variable changes over time and space, then separate models ought to be run for each period or area. This was precisely the approach adopted by the World Bank rent control project, to assess the effect of rent control under different rent control, market conditions, culture and inflation regimes (Malpezzi, *et al.* 1988). One model for the whole world would clearly have been inappropriate!

Figure 9.5 illustrates heteroskedasticity, the positive relationship between the error variance and the independent variable. The problem with this situation is that observations with high values of X provide less precise information on where the true regression line lies. A few additional positive or negative errors where X is large would bias the regression line. Heteroskedasticity can be controlled by using weighted least-squares or some other technique (Wonnacott and Wonnacott, 1970), which discounts these unrealistic observations and places more emphasis on observations where the error term is smaller.

If the error term (u) is correlated with one or more of its previous values in time $u-1$, $u-2$, or contiguous values in space, then the coefficient and intercept terms will be biased. Most econometric texts concentrate on serial correlation in the error term in time (Figure 9.6), with methods of correcting this or transforming the data; for example, regression of first differences, so that the assumptions of OLS are satisfied. Many housing studies, however, are based on spatial units: expenditure on public sector housing by administrative area, houses constructed within given areal boundaries, homelessness by cities, and so on. Spatial autocorrelation occurs when the residuals from models explaining such dependent variables are correlated with each other in

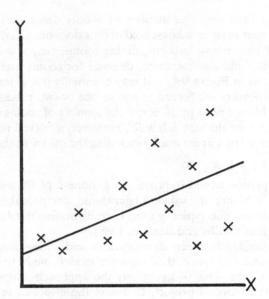

Figure 9.5 Heteroskedasticity

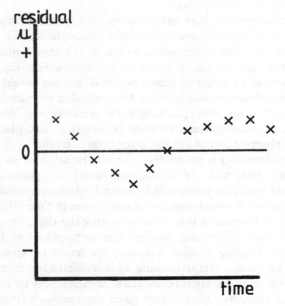

Figure 9.6 Correlated error (residual) term

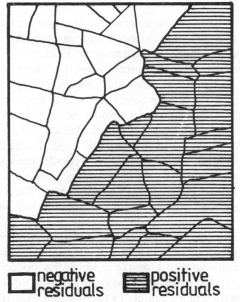

negative residuals   positive residuals

Figure 9.7 Spatially correlated error (residual) term

space. For example, instead of being random, all the positive residuals are contiguous to one another (Figure 9.7). Tests for spatial autocorrelation can be undertaken, although such tests are not simple (Cliff and Ord, 1973; Barber, 1988). The presence of spatial auto-correlation can imply (1) that a variable has been omitted from the model which explains the geographic variation in the dependent variable (2) the presence of non-linear relationships between the dependent and independent variables (3) that the regression model should have an autoregressive structure (Cliff and Ord, 1981). Thus, for example, plotting the residuals on a map may suggest additional variables to explain the residual pattern. Of course additional problems may arise in regression analysis before this stage is reached, when the dependent variable is measured with reference to areal units. Two problems can alter the sign and size of the correlation and regression coefficients. First, the scale or aggregation problem may result in the relationship between X and Y varying according to the size and shape of the spatial unit chosen for analysis. Second, even if the same number, size and shape of areal units is used (for example, one kilometre squares), different partitions of space by kilometre squares can yield different correlation and regression coefficients for the same data (Barber, 1988).

OLS assumes that the error term (u) is associated only with the dependent variable (Y). The error term can include both errors in measuring Y and any stochastic element in Y. However, independent variables (X) are also often measured with error (they may be derived from a sample) and such variables are also often subject to stochastic disturbance. If the error term is correlated with a regressor, OLS will be biased. Where the relative error in Y and X is unknown, the upper and lower bounds of the coefficient can be ascertained by regressing X on Y and Y on X (Wonnacott and Wonnacott, 1970; Byatt, Holmans and Laidler, 1973). Where the error in X is actually known, this can be explicitly incorporated into the regression analysis, and a 'true' regression coefficient calculated (Thomson and Willis, 1986). Such errors-in-variables least squares results might differ substantially from OLS estimates.

Multicollinearity concerns the common variation of two exogenous variables. With multicollinearity there is little variation unique to each variable. This is equivalent to a sample in which the independent variable does not vary very much, which means OLS has little information to use in calculating coefficient estimates. In the limiting case (Figure 9.8), in which observations on X are concentrated into a single value of X, then the coefficient (b) cannot be determined at all: any number of different

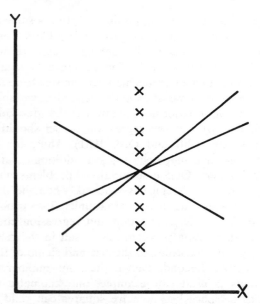

Figure 9.8 Degenerate regression because of no spread or variation in the independent variable

sloped lines pass through the mean of X and Y, fitting equally well. When two (or more) explanatory variables are highly intercorrelated, it becomes difficult to disentangle the separate effects of each of the exogenous variables on the dependent variable. The allocation of a common explanation between the two regressors is unknown, and there is thus uncertainty as to which variable deserves credit for the explained variation in the dependent variable, and hence the true value of the coefficients being estimated. However, for predictive purposes multicollinearity does not matter provided no attempt is made to predict for values of X_1 and X_2 removed from their line of collinearity. But for structural and analytical questions, the relationship of Y to X_1 or X_2 cannot be sensibly investigated when multicollinearity is present. Multicollinearity can be avoided in a number of ways, for example, by dropping from the equation one of two variables which are highly correlated with each other (say with a correlation coefficient >0.8); by deriving a composite index of such variables through principal components analysis; or by using ridge regression (available in SAS 1986).

LIMITED DEPENDENT AND QUALITATIVE VARIABLES

Special problems arise when (1) the dependent variable or one or more of the independent variables is measured not on a cardinal scale, but on an ordinal or nominal scale, or (2) when the observations are drawn from a disproportionate sample. Such problems often arise in housing studies.

Consider first dummy variables which can be used in an OLS regression, where an independent variable takes on the value of either zero or one. For example, housing consumption or expenditure (the continuous dependent variable) may be hypothesised to be a function of income, tenure group (owner occupier, private rented, or rented from the government), and ethnic group. Tenure group measures property rights over the house; while one ethnic group may be hypothesised to prefer more housing, *ceteris paribus*, and the other ethnic group less. Tenure and ethnic group are categorical variables, which must be represented by a variable of value 1 or 0, or a dummy variable. The dummy variable measures the effect of being in a particular class relative to not being in that class. The implicit assumption is that the regression lines for the different groups differ only in the intercept term, that is they have the same slope coefficients. The coefficient of the dummy variable measures the differences in the two intercept terms. There should always be one less dummy than the number of groupings by that

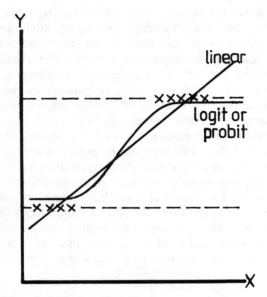

Figure 9.9 Discrete choice regression

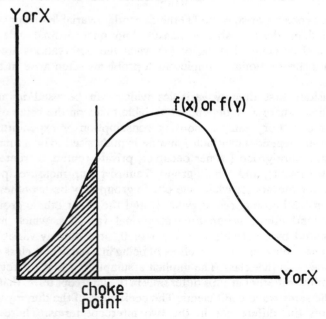

Figure 9.10 Censored and truncated distributions

category, because the constant term is the intercept for the base group and the coefficients of the dummy variables measure differences in intercepts. Thus, if there are two ethnic groups, there is one dummy variable in the equation; which takes on a value of 0 if the household belongs to ethnic group A, and a value of 1 if the household belongs to ethnic group B. For three tenure groups, there would be two dummy variables. The base case, say government housing, is the omitted category. The two dummy variables would record whether the household was in a private rented house (0 or 1) and whether the household was in an owner occupier house (0 or 1). If the intercept term (a) in the equation, and all three dummies were included in the equation, perfect multicollinearity would be introduced. Hence one category must be omitted.

In some models the dependent variable might be a dummy variable. Such a dummy variable can take on two or more values, but the simplest case occurs where it takes on only two values, $Y = 0$ or $Y = 1$. Typical examples might be a person owns a home ($Y = 1$) or he does not ($Y = 0$), a housing mortgage is granted ($Y = 1$) or it is rejected ($Y = 0$), a new firm fails and goes bankrupt ($Y = 1$) or it succeeds ($Y = 0$). Such events can be analysed with a linear probability model, which is closely related to the linear discriminant function used to analyse tenure choice (Chapter 8). However, in a linear probability model, predicted values can lie outside the interval (0,1), a situation which can also occur when using an OLS model where the dependent variable is a percentage between 0 and 100. The estimated probabilities can be squeezed inside the 0–1 interval by a cumulative normal function (probit model) and a logistic function (logit model) (Figure 9.9). The probit and logit models often produce similar results when the coefficients are normalised (Maddala, 1988). These models have been employed to assess discrimination in housing loan markets (Maddala and Trost, 1982); in analysing tenure choice in housing markets (Friedman et al, 1988), in estimating willingness to pay for water (in categorical ranges for yes/no responses to discrete prices) and the probability of the use of new water sources if different prices were charged (Chapter 11), and to assess the factors instrumental in influencing condition of houses later in this chapter.

In some samples no information may exist on a certain section of the population. For example, in analysing the effect of income (X) on housing consumption (Y), it may be decided to omit households living in squatter settlements, that is houses with a low capital value, in the informal sector without legal status. In this sense the sample is truncated (Figure 9.10): house values below a certain threshold level are omitted, and a random sample selected from the remainder of the housing stock. Housing allowance experiments in the USA have excluded households above a certain income level. Again this is an example of a censored sample where the sample is truncated on the basis of an independent variable.

With such a set of observations, OLS will give a biased estimate of the effect on housing consumption of income. The regression equation

$$Y = a + bX_i + u_i$$

where Y_i is only observed if $Y_i \geq c$; and c is some constant or choke value. Since $bX_i + u_i \geq c$ or $u_i \geq c - bX_i$, the error term is not equal to zero, it will be a function of X_i. Hence an OLS estimate of the coefficient (b) will be upward biased, giving an over-estimate of the true effect. A truncated maximum likelihood estimator and probit regression can correct for truncated and censored samples (Maddala, 1983).

The following example analysing occupancy rates demonstrates, as does the econometrics chapter (10) on the impact of rent control, the use of OLS multiple regression with dummy variables and transformations. The house condition example (along with Chapter 11) illustrates the use of regression analysis where the dependent variable is discrete.

MODEL OF OCCUPANCY RATES

The area to which this model applies is Kumasi, one housing market area in the Asante cultural region which dominates the central tropical forest area of Ghana. Alongside data from other African countries (e.g. in Peil and Sada, 1984), housing in Kumasi is overcrowded: there is a relatively high occupancy rate or number of people per room. For rented housing, the average number of persons per room is 3.5, with a median of 3.0. In 1986, household size ranged from one person households to one sampled household with 52 persons. While extended family households are encountered, it is noticeable that there is a very gradual fade in the frequency of these families in households above six persons. For family housers (those related to the owner of the house or themselves being a joint owner of the house inherited under the family lineage), the mean number of people per room is 3.2, with a median of 2.5; and for owners the respective figures are 2.7 and 2.0.

From survey information on the number of people who formed the household and the number of rooms they occupied, a continuous variable, occupancy rate, can be defined. An OLS regression model was used to explain variations in occupancy rates for households across all tenure sectors. The model took the form

$$Y_i = a + b_1 X_{1i} + b_2 X_{2i} + ... + b_k X_{ki} + u_i, \quad i = 1, 2, ... n$$

where the explanatory variables $(X_1, ..., X_k)$ are either continuous or dummy variables. The independent variables, grouped under three headings, demographic, economic and property right variables, hypothesised as explaining or influencing household occupancy rates, are listed in Table 9.1, together with the rationale behind each variable. The dependent variable is the number of people per room, that is, co-resident household size divided by the number of rooms occupied.

Total household size is an obvious variable which affects occupancy rates per room. This was omitted as an explanatory variable in the right-hand side of the model, since it is the numerator of the fraction used to derive the value on the left hand side of the equation.

The OLS results are presented in Table 9.2; and are based on 799 households for which information was available on all the variables included in the model, out of the 1414 households in the sample. Only statistically significant variables were included in the model, and these produced an adjusted R^2 of 0.7743. Expressing three of the independent variables in logarithmic form not only increased the goodness of fit of the model, but also eliminated heteroskedasticity in the residuals.

Almost all the variables in the model had the right expected sign. A major variable in explaining occupancy rates was income per capita, where the relationship was negative: as income per capita increases occupancy rates decline. Rent paid relative to income was used as a proxy for the preference for housing or degree of substitution of housing for other goods. The negative coefficient on this variable reveals that occupancy rates will be lower if households have a preference for more housing *vis à vis* other goods. Conversely, the number of children in the household is positively related to occupancy and contributes directly to higher occupancy rates per room.

Family housers are associated with lower occupancy rates, while households occupying units outside rent control also have lower occupancy rates. Rent control appears to increase occupancy rates, probably by reducing the supply of units to rent, and also through inhibiting households changing residence to match their demographic requirements with house structure and space, because of economic rent seeking behaviour. Economic rent seeking can be large in developing countries (Krueger, 1974), and it often dissipates the utility or benefit of those receiving it or results in the additional expenditure of resources to

Table 9.1   Explanatory variables of household occupancy or overcrowding
(variables and rationale)

Demographic variables

1.   Number of children: contributes directly to high occupancy per room.
2.   Female head of household (dummy): more likely to be associated with
      overcrowding because of inability to afford more rooms.
3.   Akan household (dummy): non-akans more likely to require more rooms
      on religious and cultural grounds to separate male and female sleeping
      quarters.
4.   Co-resident household (dummy): households which are not co-resident
      (man and wife live in separate houses) will have more rooms and therefore
      a lower occupancy rate per room. Non-coresident households are
      traditional Ghanaian.
5.   Family type (dummy): cultural obligations to provide accommodation for
      relatives would suggest extended family households should be more
      overcrowded.

Economic variables

6.   Number of workers in household: generates more income and therefore
      demand for more privacy and space.
7.   Income per capita: higher income permits more accommodation (rooms)
      to be rented.
8.   Rent to income ratio: expresses household preference for housing *vis à vis*
      other goods.

Property rights variables

9.   Dwelling outside rent control: rent control is likely to generate economic
      rent seeking behaviour thus inhibiting movement of households to equate
      demographic requirements with housing structure. Could result in either
      under or overcrowding.
10.   Length of residence in the house: households less inclined to move have
      higher occupancy rates. The mover-stayer dichotomy in housing (and
      migration and labour mobility in general) has long been recognised.
      Such markets are characterised by a few households who frequently move
      and others who experience long durations at the same residence
      (Bartholomew, 1967; Willis, 1974). The latter households who do not
      adjust housing to increases in family size are likely to experience greater
      occupancy rates per room.
11.   Tenure type (dummy): family housers who have legal rights over tenants
      are expected *a priori* to result in lower occupancy rates.

achieve it (McKenzie and Tullock, 1981; Buchanan, 1983). Once outside
rent control, household requirements are more easily matched with
housing space. Length of residence is also associated with higher
occupancy rates.

    Other factors contributing to increased occupancy rates are female

Table 9.2 OLS regression of household occupancy rate.

Dependent variable: number of persons per room in household
Degrees of freedom: 799

R-square: 0.7774   F: 250.242
Adj R-sq: 0.7743   Prob > F: 0.0001

Variable	Parameter estimate	Standard error	T statistic	Prob > /T/
Intercept	2.6831	0.2636	10.17	0.0001
Log number of children	0.4135	0.0246	16.78	0.0001
Female head of household	0.0775	0.0277	2.79	0.0053
Co-resident household	0.3378	0.0453	7.45	0.0001
Nuclear family household	0.6265	0.0421	14.88	0.0001
Extended family household	0.6128	0.0478	12.80	0.0001
Log number of workers	0.1488	0.0311	4.77	0.0001
Log income per capita	−0.3364	0.0281	−11.93	0.0001
Rent/income ratio	−2.7596	0.2820	−9.78	0.0001
Unit outside rent control	−0.2178	0.0259	−8.40	0.0001
Log length of residence	0.0481	0.0127	3.76	0.0002
Family houser household	−0.5021	0.1201	−4.18	0.0001

headed households; co-resident households (see below); and the nuclear family and extended family households. It may be assumed that female headed households have less ability to pay for housing and thus consume less. However, two factors should be mentioned to qualify this. First, many of the richest traders in Kumasi are women, thus the woman of a household may be much more economically powerful than the man. Second, Asante traditional marriage is non-coresidential: it is traditional for a man to live separately from his wife in a different house, where she joins him for conjugal relations and to which she may send some of his meals. Thus, many seemingly female headed households are not without the additional support of a man and may be able to afford the same quantity of housing as a male headed household. On the other hand, the splitting of a man's accommodation from the rest of the nuclear family requires more housing to be consumed leaving, therefore, fewer resources from which to house the woman and her children. Where Asante households have such an arrangement it obviously reduces occupancy rates to some extent; and conversely co-residential marriages increase them. Extended family household arrangements also contribute to higher occupancy rates; since where space is available there is a cultural obligation to accommodate extended family members, even where households have only one room to accommodate themselves and two children, they may be joined by the wife's sister or other relative.

MODEL OF HOUSE CONDITION

House condition is difficult to assess. It is largely a qualitative variable in respect of house structure. The complete lack of empirical data on improvements, conversions, and maintenance expenditure on existing units which affects conditions – and also which prevents deterioration and depreciation, thus reducing gross and net supply losses (in quantity and quality) from the existing housing stock – adds to the problem.

Some cross-section data was available on the condition of houses in the 1986 housing survey of Kumasi (Malpezzi, Tipple and Willis, 1990) consisting of categorical data on whether there were cracks in the wall surface of the house, the condition of the plaster, the courtyard surface, the state of the gutter (a concrete structure running around the base of the house for drainage), and the seriousness of these defects. Condition data also existed on external house features such as the condition and quality of outside surfaces, and whether there was standing water, vegetation, and rubbish lying around (each of which may encourage mosquitoes and other disease vectors).

The condition of Kumasi housing was modelled by treating the dependent variable (i.e. house condition) as a dummy variable which could take on two or more values. In our case, the dummy dependent variable is taken to be dichotomous, where $y = 1$ if the house is in good condition, and 0 if the house is in poor condition. The dichotomous dummy dependent variable was derived from an amalgam of various data on the physical condition of the house, from a survey of individual houses in Kumasi in 1986. The independent variables hypothesised as influencing housing conditions are listed in Table 9.3, together with the rationale behind each variable. In the event, the impact of rent control on house condition could not be evaluated, since parts of all houses were subject to rent control, even though some individual households within houses were not.

A conditional probability model was used to analyse the data and predict house condition. Conditional probability models can be used to estimate the relationship between a set of attributes describing an entity and the probability that the entity will end up in a given final state. The simplest form is the linear probability model

$$Y_i = a_0 + bX_i + e_i$$

where $Y = 1$ if the house is in good condition

Table 9.3 Explanatory variables of house condition (variables and rationale)

1. *Year of construction*
    As a house ages, it depreciates and its condition deteriorates.
2. *Swish wall material (dummy)*
    Mud (swish) houses constructed of laterite clay alone without any cement
    additive are likely to deteriorate and generate major cracks and faults to a
    greater extent than cement structures, especially where foundations are not
    well built or roof overhang is insufficient to protect the walls from rainwater
    damage.
3. *Compound house (dummy)*
4. *Multi-storey house (dummy)*
    A multi-storey house is likely to incur higher maintenance costs than a
    single-storey house, and consequently multi-storey houses might be
    expected to be in poor condition. On the other hand multi-storey houses
    are built of superior materials so are unlikely to deteriorate in condition as
    quickly as other houses. Thus the sign of the relationship between this
    variable and the dependent variable is indeterminate *a priori*.
5. *Nearness to a road*
    A continuous (distance) variable standing as a proxy for whether the area
    was planned and the likelihood of water in the house. The further from a
    recognised road the greater the probability that the area is unplanned and
    the lower the probability that mains water is provided in the house. This
    would reduce rental value and therefore increases the likelihood of poor
    maintenance and non-improvement of the dwelling.
6. *Owner on premises (dummy)*
    Owners resident on the premises are more likely to maintain the house and
    also to respond to tenant pressure for repairs.
7. *Family house (dummy)*
    This variable stands as a proxy for the prisoner's dilemma problem. Multiple
    owners might be hypothesised as likely to lead to problems of co-ordinating
    maintenance.
8. *Akan area*
    The percentage the local Akan ethnic group in an area stands as a proxy for
    housing externalities and service conditions and provisions. Service
    provision (for example, water, paved roads, street lights, etc.) are much
    worse in non-Akan areas. Thus more desirable neighbourhoods are likely
    to encourage housing maintenance.
9. *Number of households in house*
    This variable stands as a proxy for greater wear and tear on fabric and
    excessive use of services in the house. The greater the number of house-
    holds in a house, the greater the likelihood that the house is in poor
    condition.
10. *Income (dummy)*
    Above and below average per capita income per house. Expressed as a
    dummy variable since the number of persons in a household varied, from 1
    to 52 (median = 5). The number of households per house varied too. The
    greater the per capita income, the greater the ability to undertake mainten-
    ance (if the landlord does not in rental housing).

0 if the house is in poor condition

X = value of an attribute for house i

e_i = independently distributed random variable with mean of zero.

This regression equation describes the probability that a house will end up in a certain condition, given information about attributes of the house, household or neighbourhood. Although the linear probability model is appealing because of its simplicity, there are a number of problems associated with it. Problems range from the generation of (1) probabilities which lie outside the 0–1 limits; (2) negative probabilities; (3) an error term which is not normally distributed leading to problems with the application of the usual tests of significance; and (4) heteroskedasticity in the error term leading to inefficient estimates. While weighted least squares can correct for heteroskedasticity, probabilities may still lie outside the 0–1 range.

An alternative probability model is therefore desirable which will constrain all predictions within a 0–1 limit. The logit model transforms the values of the attribute X to a probability which lies in the range 0–1, the model being based on the cumulative logistic probability function

$$P_i = F(a + bX) = \frac{1}{1 + e^{-(a + bX_i)}}$$

where P_i = probability of the house being in good condition

X_i = attributes of the house and household

e = natural logarithm

The results of the logit regression of house condition are presented in Tables 9.4 and 9.5. Table 9.4 shows the influence of variables, through the logit regression coefficients, in determining house condition. The three significant variables determining house condition are distance from road, owner living on premises, and number of storeys in the building. House condition is negatively related to distance from a road and positively related to the owner living on the premises, confirming *a priori* hypotheses outlined in Table 9.3. In the specification of variables, multistorey houses were assigned a value of 0 and single storey houses a value of 1. Thus the negative coefficient on storeys indicates that single storey dwellings are more likely to be in poor condition than multistorey dwellings, attesting to the superior materials and construction of the latter type of house.

Other variables included in the model did not prove to be statistically significant in determining house condition, although the signs on the variables were generally in the right direction. Compound houses are

Table 9.4 Probability of 'better' house condition: logit function for complete data/variables

Dependent variable: house condition
Number of observations: 185
Number in poor condition: 95
Number in good condition: 90
Observations deleted due to missing values: 105
Chi-square = 39.09

Variable	Beta	Standard error	Chi-square	P	R
Intercept	2.7529	1.4968	3.38	0.0659	
Compound	−0.3498	0.4601	0.58	0.4471	0.000
Swish	−8.7339	20.4906	0.18	0.6699	0.000
Distance to road	−0.0041	0.0017	6.00	0.0143	−0.125
Owner on premises	0.6322	0.3672	2.96	0.0852	0.061
Family house	0.1157	0.4138	0.08	0.7797	0.000
Akan	−1.5822	1.4769	1.15	0.2840	0.000
Number of households in house	−0.0417	0.0335	1.55	0.2131	0.000
Storeys	−1.1215	0.4904	5.23	0.0222	−0.112
Income	−0.1526	0.3609	0.18	0.6725	0.000

Table 9.5 Predictive values from the logit regression policy capturing equation model, compared to actual (true) values

	Predicted classification		
	Negative	Positive	Total
True classification			
Negative	61	34	95
Positive	21	69	90
Total	82	103	185

Sensitivity: 76.7 per cent
Specificity: 64.2 per cent
Correct: 70.3 per cent

more likely to be in worse condition than other types. Swish (mud) construction also contributes to poor condition, with the coefficient here being quite large although not statistically significant. Contrary to *a priori* expectations, however, family houses were more likely than others to be in good condition, *ceteris paribus*; suggesting strategic behaviour, free rider and prisoners' dilemma problems were not important in house maintenance. Prisoner's dilemma is a situation in which it pays several economic agents individually to behave in a particular way, even though it would pay them as a group to behave in some other way (Davis and Whinston, 1961; Willis, 1980). It may be that the cultural and family obligations of maintaining the family home outweigh any free rider possibilities. Indeed, Becker (1981: 32n, 118, 237) argues that family members monitor one another to protect against shirking, malfeasance and other 'moral hazards' of insurance by kin, including non-payment of debts. Antisocial acts by individuals are punished by family members. On the other hand, it may be that the generally poor maintenance level in Kumasi in general was not significantly exceeded by family houses. The more heavily Akan dominated houses and areas were associated with poor housing conditions, suggesting that, in rental housing, where cultural and family obligations were absent, houses deteriorated. This is consistent with observations that, in Akan culture, little kudos attaches to maintaining houses (Tipple, 1983). Rent controls merely reinforce this preference.

Increases in the number of people and families in a house was associated with worse conditions, as expected. However, above average per capita income in a house was not positively associated with better conditions, contrary to prior expectations, although the coefficient was not statistically significant. Age of house was omitted from the model on statistical grounds since it reduced the number of observations available: house age data were frequently missing or unreported.

The allocation of individual houses on the basis of this model to worse or better condition classifications respectively, are presented in Table 9.5. Ninety-five houses were actually in worse condition and 90 were in better condition than average. The model predicted 82 and 103 respectively in these categories. Thus, this model, trying to capture the outcomes of individual decisions, performed much better than random, achieving a 70.3 per cent overall correct classification of houses.

Nevertheless, the classification error in this model is higher than that in logit models used in other fields such as those used to predict company (firm) failure (where misclassification overall is typically 10 per cent to 20 per cent) (see Storey, Keasey, Watson and Wynarczyk, 1987), or housing tenure (see Friedman, Jimenez and Mayo, 1988). Logistic models have been extensively used to analyse company failure, compared to survival or success, and a considerable body of literature has been built up in this field. The independent variables used in these models are measured with considerable accuracy and consistency being derived from company accounts data. Moreover the 'true' classification of observations is more 'objective' in this field: both company failure and house tenure are defined in law! In comparison the 'true' classification of house condition is much more imprecise and subjective, both on the part of what the enumerator perceives and in combining the variables to form a composite house condition index to indicate better or worse condition. Nevertheless, where logit models have been used to predict company failure in the future, say two, three, or five years in advance, correct classification rates are much lower (typically around 70 per cent). Thus, given the length of time required for a house to deteriorate, and deterioration which is a consequence of variables present in the past (which may not be the same current factors observed now associated with the house and its occupants), a correct classification rate of 70 per cent can be regarded as a reasonable model.

Within the overall correct classification rate, the positive and negative hit rate varied. The sensitivity rate, or the proportion of all houses which were actually in better condition and were predicted as such by the model, was 76.7 per cent. Thus, given a house is in good condition, there is a 76.7 per cent chance that it would be predicted to be in good condition according to the model. The model performed less well in confining (predicting) worse condition houses to those actually in this category: hence the specificity rate was only 64.2 per cent. The model can also be judged on its predictive ability. Thus, the predictive value positive is 67.0 per cent: given the house is predicted to be in good condition, there is a 67.0 per cent chance of its actually being in good condition. Similarly, given a house predicted to be in poor condition, there is a 74.4 per cent chance (predictive value negative) of its actually being in poor condition.

Overall this policy capturing equation model of house condition is somewhat short of 100 per cent diagnostic ability. But complete infallibility is never achieved, neither in a policy capturing or deductive model nor in terms of intuitive judgement by an individual 'expert'. Thus, a model such as this, naive as it currently is, would still out-perform an 'expert' in assigning unseen houses to a better or worse condition classification on the basis of variables influencing house condition. The power of policy capturing models to outperform experts, and the inaccuracy of expert opinion and intuitive judgement alone, has been deomonstrated in a number of fields from psychology (Meehl, 1954; Maniscalco *et al.* 1980; Kleinmuntz and Szucko, 1984), labour selection (Dawes, 1980), and criminology (Glaser, 1985), to medicine (de Dombal *et al.*, 1974; de Dombal, 1984; McGoogan, 1984).

CONCLUSIONS

Regression analysis can be a useful tool in evaluating qualitative as well as quantitative issues. It aids in structuring questions about the relationships between variables, and the efficiency, consistency and unbiasedness of analytical and predictive models. Where tasks or problems can be structured in a form amenable to system aided judgements, statistical models will always outperform experts in diagnostic accuracy, even when the cues or coefficients in the regression model are the weights attached by experts themselves to the variables explaining the dependent variable. Sheer ignorance, computer phobia, dehumanising flavour, threat of unemployment and replacement by machines, and the professional ethic, are factors considered by Meehl (1986) to explain the ubiquity and recalcitrance of irrationality in the conduct of human affairs, and the continued practice of using intuitive judgements and experts' opinions, when data based aids or even 'bootstrapping' would prove more accurate.

ACKNOWLEDGEMENTS

The author gratefully acknowledges a grant by the Leverhulme Trust in support of this research and the World Bank for permission to use the 1986 Kumasi data set. The views and opinions expressed, however, are those of the author alone.

10 Econometric analysis

Measuring the impact of rent controls in urban housing markets

Raymond Struyk and Margery A. Turner

INTRODUCTION

Controls on the rents that landlords may charge their tenants are quite common in developing countries, as well as in the United States and the industrialised nations of western Europe. Indeed, it is much easier to think of countries that have some form of rent controls than of those that do not. In many instances, controls are accompanied by regulations protecting tenants from eviction. There is, however, little uniformity across countries in rent controls and accompanying protective legislation or in the administration of these regulations. Thus, it is difficult to generalise from one country to another about the likely effects of rent control on housing market outcomes.[1]

While the primary effect of rent controls is to reduce the cost of rental housing for its occupants, reduced rents may produce a number of important indirect impacts.

1. On housing supply

By reducing rent levels, and possibly appreciation rates as well, rent controls may reduce the profitability of rental housing and cause a reduction in the production of rental housing or a withdrawal of units from the rental market.

Profitability is also the primary determinant of investment in unit maintenance. Lower profits after some point may result in lower maintenance, lower unit quality, and shorter unit lives.

2. On housing demand

Changes in the relative price of owner and renter occupied housing may change the tenure choice of some households.

Reduced rents to long-term renters may make relocation very expensive, thus discouraging residential and possibly labour mobility as well (Clark and Heskin, 1982).

This chapter's primary goal is to describe one approach, which has been implemented for urban Jordan, for rigorously analysing the economic benefits accruing to tenants under a rent control regime in a developing country. In addition to estimating the average level of benefits, this analysis documents the distribution of benefits among renters with different income levels, household sizes, and length of tenure in their units. After presenting these primary findings, some of the possible indirect effects of rent controls are considered, focusing in particular on methods for analysing the effects of controls on profitability and investment in rental housing. In this case we employ information obtained both for Jordan and for Washington, DC. Interestingly, Washington's system of rent controls is quite similar to Jordan's in several important ways, including landlords' ability to raise rents to market levels whenever a unit is vacated.

The various analyses described in this chapter do not result in a single, unambiguous conclusion for or against rent controls. Instead, they dramatise the importance of conducting careful, empirical analysis for individual housing markets and regulatory regimes. More specifically, the studies described here concluded that:

1 Examining the distribution of rent control's benefits is as important as estimating their magnitude. In Jordan, for example, benefits proved to be far from universal. While lower income renters received significant benefits, a substantially larger share were enjoyed by relatively affluent renter households.
2 Estimates of landlords' profits, with and without controls, can provide important insights into the potential impacts of controls on housing investment. In Washington, DC, rent control significantly reduces cash flow, but after adjusting for expected appreciation and tax benefits, the profitability of controlled rental units compares favourably to alternative investment opportunities.

The effects of rent controls on housing availability and quality are extremely difficult to disentangle from the effects of other supply and demand factors. For example, a dramatic decline in the number of rental units in the District of Columbia is often cited as evidence of rent control's adverse impacts. Comparison with other US central cities during the same time period, however, strongly suggests that national demographic and macro-economic conditions played a more significant role in determining rental housing supply than did the local regulatory environment. On the other hand, in Jordan strong tenant protection provisions, which are part of the broader rent control regime, appear to discourage many families' working outside of Jordan to rent their homes until they return.

Almost by definition, rent controls are intended to reduce the amount tenants pay for their units. However, the magnitude of these rent savings, the extent to which they benefit renter households, and the distribution of these benefits across socio-economic groups are all difficult to measure empirically.

CONCEPTUAL FRAMEWORK

One can think of the benefits to a tenant from rent control as simply being the difference between the rent he would be paying in the same unit if rent controls were not in effect. Correspondingly, the value of rent controls could be estimated from a hedonic model, in which all of a dwelling's attributes are controlled for, and the benefit of controls is the discount associated with length of time the household had been in the unit.

Unfortunately, this approach does not provide a good estimate. Differences between such discounts and the net benefits to tenants arise because tenants may be consuming less or more housing services than they would have in the absence of controls. These disequilibrium effects presumably become greater the longer a tenant is in a unit.

To estimate net tenant benefits, we follow the procedure developed by Olsen (1972) in his analysis of rent control. Based on microeconomic theory, and employing the concept of consumer's surplus, he derives the following formula to calculate net benefits (*NB*), expressed in cash equivalent terms, from rent control for a household.[2]

$$NB = PmQm - PcQc + PmQm \, [lnPmQc - lnPmQm] \qquad [1]$$

where

Pm is the price per unit of housing services for uncontrolled units,
Qm is the quantity of services consumed by the household in an uncontrolled unit,
PmQm is the market rent for the unit selected in a market without rent controls,
Pc, *Qc*, and *PcQc* are equivalent concepts in the controlled market, and
ln denotes a natural logarithm.

Net benefits can be thought of as the amount of money sufficient to make the tenant indifferent to either staying in his/her rent controlled unit or moving to the unit of his/her choice at the market rent, ignoring the cost of finding the unit and actually relocating. The first component of [1], i.e. $PmQm - PcQc$, is the benefit accruing to the tenant from having additional money to spend on other goods as a result of living in the rent controlled unit. Because Pc is less than Pm, one might think that Qc is greater than Qm. However, Olsen (1972) argues that under rent control there is no reason for the consumer to be on his demand curve. Moreover, there are clear incentives for suppliers to reduce Qc by cutting back on maintenance and other services, for example, repainting the unit every few years. On balance, Olsen actually found reduced consumption (Qc is less than Qm) for controlled units in New York City. Hence, it is unclear *a priori* whether the second term will increase or decrease tenant benefits.[3]

It is worth noting that the calculation of consumer surplus engenders some problems. In particular, as described below, estimation of a demand function to obtain $PmQm$ may be fraught with the problem of mismeasuring market rents. Less severe, but of potential difficulty, is the requirement for a well-specified hedonic model to measure $PmQc$.

MEASURING THE INPUTS

To compute net benefits for each household using [1], we need to know (a) its housing demand under uncontrolled conditions ($PmQm$); (b) its housing consumption in the controlled market ($PcQc$); and, (c) the market rent of the unit occupied under rent control ($PmQc$).

Before detailing the computational procedures, we should say a word about the data set employed in this study. These data come from a national housing survey, employing a clustered random sample design, conducted in Jordan in the summer of 1985. The total sample size was about 2300.[4] It collected detailed information on the current housing unit, length of tenure, the occupants and their economic activity, and the household's expenditures on housing. The sample for greater Amman included a total of 765 observations (both occupied and vacant dwellings), of which 257 were occupied rental units. The combined sample for other urban areas was 793 dwellings, of which 177 were occupied rental units.

First, $PcQc$ is the rent reported by the respondent for the unit presently occupied. Second, for the market rent of the current unit ($PmQc$), two options were available. One was to use the response to the question in the survey for the household's estimate 'for the present market rent of a similar unit in this neighbourhood'. The alternative was to employ a hedonic model estimated for rental units by substituting the values for the unit's attributes into the model but setting the length of time in the unit to zero. Ultimately, it was decided to use both approaches. The household's estimate is important as it, in effect, reveals the size of the benefit which they believe they are obtaining from rent controls.

The hedonic estimate provides a somewhat 'more objective' measure.[5] As shown in the results for the hedonic indices given in Table 10.1, the data set included a rich set of variables describing the unit; and reasonably good estimates were obtained in the sense of the degree of variance in rents explained and the signs of the coefficients. The variable for the length of time the household had occupied the unit was highly significant in the models for both Amman and other urban areas.

For the market rent of the unit that the household would have occupied in the absence of rent controls ($PmQm$), a simple demand function was estimated in which the samples employed for Amman and other urban areas included only those renters who had moved into their units in the past three years, i.e. recently moving households. The independent variables are household income, a dummy variable indicating that the household is receiving remittance income (which might influence housing consumption in either direction[6]) the number of persons in the household, a set of dummy variables indicating the amount of savings the household reported in terms of monthly income equivalents, and some location variables.[7]

The estimated demand functions are presented in Table 10.2. As shown, household income is the only variable statistically significant in both models. The income elasticity of demand, evaluated at the means, is 0.2 for Amman and 0.4 for other urban areas. In other urban areas, two of the variables for accumulated savings were significant, and they suggest a quite irregular effect of savings on housing demand. (The omitted savings category is savings equivalent to more than six months of income.) In Amman, household size exerts significant pressure on housing consumption, with each additional household member increasing rents by about JD3.20 or about 7 per cent of the average rent.[8]

Lastly, of the locational variables included, only the newly developing and high quality Wadi El Seir area within Greater Amman was found to have the differential effect of a distinct submarket. And its effect is very large indeed. Still, this value does not seem out of line given the new luxury housing and the prestige value of the location, qualities for which a subset of renters appear quite willing to extend themselves.

Table 10.1   Hedonic indices for Amman and other urban areas: dependent
variable: contract rent

Independent variables	Amman	Other urban
Type of unit		
Traditional detached (1=yes)	–5.30	–4.46
Apartment (1 = yes)	–0.70	a
Villa (1 = yes)	6.32	a
Unit Size (sq. m)	0.35**	0.14**
Unit Size squared	–4.81E–05	
Separate kitchen (1 = yes)	–2.26	–1.96
Separate bathroom (1 = yes)	5.55*	1.72
Separate WC (1 = yes)	–3.14	–10.39
Drinking water		
Piped water	–5.30	a
Public tap	a	a
Water tanker/vendor	a	–21.30
Waste water		
Public sewer	a	–4.99
Septic tank	1.70	–1.80
Lighting by electricity (1 = yes)	–2.32	0.37
Type of heating		
Central	23.69**	a
Natural gas	–8.45	–6.00
Kerosene	a	a
Wood/charcoal	a	a
Roof is concrete metal (1 = yes)	–10.55	4.62
Good access		
Grocery store	–4.33	a
Primary school	0.09	a
Police station	0.73	a
Post office	–1.70	a
Health centre	5.06*	a
Satisfied with neighbourhood (1 = yes)	–4.88	3.12
Time in unit (years)	–2.30**	–0.93**
Time in unit squared	0.03**	9.48E–03
Rental contract (1 = yes)	4.26	7.40**
Utilities included in rent (1 = yes)		
Water	–7.25	–5.77
Electricity	8.68	7.80
Central heating	–4.67	a
Offered money to leave (1 = yes)	3.28	0.48
Owner lives in building (1 = yes)	–2.63	–2.99
Type of zoning (1 = yes)		
A or B	8.68*	5.11
C or D	1.86	–0.70
Agriculture, industrial	–61.66**	–10.86*
Residential with other use (1 = yes)	9.94	4.86

Table 10.1 Continued

Independent variables	Amman	Other urban
Wall material (1 = yes)		
Finished stone	37.86**	9.65
Concrete or cement block	27.14*	4.08
Year unit built	0.05	0.09
No. units in structure	0.40	−0.08E−03
No. floors in unit	3.45**	2.91
Persons per room	−1.52	−0.74
Location		
Sweileh	−6.41	
Salhiet Al Abed	a	
Wadi El Seir	65.19**	
Madaba		5.26
Zerka		−4.03
Ruseifeh		−8.36
Irbid		−6.10
Mukheim Alhuson		−16.31
Constant	12.65	23.79
R squared	0.82	0.64
R squared (adj)	0.77	0.44
SEE	16.74	13.22
F	17.32	3.23
d.f.	146	64

Notes:
** Significant at 0.05 level or higher.
* Significant at 0.10 to 0.05 level.

RESULTS

Using the estimates of $PmQm$, $PmQc$, and $PcQc$, net benefits (NB) were computed for each renter household. The mean values for NB and the other factors are shown in Table 10.3 for each of the two variants used for $PmQc$, i.e. the estimated market rent of the unit currently occupied.

As can be seen, there is considerable variation among the estimated mean net benefits, depending on which area and which assumption for $PmQc$ is used. The extreme values are for other urban areas, where the mean net benefits as a percentage of current rent ($PcQc$) range from 60 per cent for the perceived market rent to 7 per cent for the rent estimated with the hedonic function. For Amman, the parallel figures are 13 per cent and 27 per cent. The larger variance in the figures for other urban areas is caused primarily by a much greater difference between the estimates of $PmQc$ and $PcQc$ for these areas than for

Table 10.2 Estimated housing demand functions[a] dependent variable: contract rent

	Amman	Other Urban
Constant	24.43**	1.96
Income (JD per month)	0.04**	0.08**
Remittance income (1=yes)	13.35	3.16
Household savings, in month-expenditure equivalent		
Less than 1 month	–4.41	17.50**
1–2 months	1.31	–3.25
3–6 months	2.33	17.92**
Persons in household	3.19**	–0.15
Location		
Sweileh	–3.71	
Wadi El Seir	93.68**	
Irbid		5.72
Madaba		–3.26
Zerka		4.30
Ruseifah		–5.66
R^2	0.47	0.36
R (adj)	0.44	0.29
F	13.15	5.61
(sign.)	(0.000)	(0.000)
d.f.	117	102
means	50.61	35.07
Income elasticity[b]	0.19	0.43

a   Sample limited to those moving into unit in last 3 years.
b   Evaluated at the means.
** Significant at 0.05 level or higher.
*   Significant at 0.10 to 0.05 level.

Table 10.3 Alternative estimates of net benefits from rent controls

	NB	PmQm	PmQc	PcQc
Amman				
Perceived market rent	5.74	53.75	56.88	43.38
Estimated market rent	11.59	53.75	61.13	43.38
Other urban				
Perceived market rent	19.76	33.71	61.37	32.65
Estimated market rent	2.34	33.71	37.25	32.65

Amman. Households in other urban areas apparently believe that, on average, market rents for the equivalent of their current unit would be almost double their current rent, whereas those in Amman estimate the difference as equivalent to about a one-third rent increase, on average. The authors do not know what accounts for these differences in perceptions between Amman and other urban areas.

Taking the estimate of net benefits (*NB*) made using the hedonic estimate of the market rent for the current unit as the more reliable, it may be concluded that rent controls are providing substantial average benefits in Amman; equivalent to 27 per cent of current rents and 4 per cent of the mean income of renters. In other urban areas, in contrast, benefits would only be equivalent to 7 per cent of current rents and a little over 1 per cent of income. The lower benefit rate in other urban areas in part results from lower tenure periods among the households included in our sample. In Amman, 34 per cent of our sample of renters moved into their units in the last two years, while in other urban areas about 50 per cent had; at the other extreme, in Amman, 30 per cent had lived in their unit for over ten years, while the similar figure for other urban areas is only about 11 per cent. The lower benefits in other urban areas may also reflect less rigorous enforcement of rent controls outside of Amman. At least as important as the mean benefits from rent control, however, is the distribution of these benefits among different groups of households. Of particular interest to us are the variations in benefits among income groups, the rent levels paid, household size, and the length of time a household has occupied its unit. To summarise these patterns, we have estimated a regression model in which the dependent variable is estimated net benefits (*NB*) and the independent variables are income, rent, length of time in unit, and household size.

The estimated models are presented in Table 10.4. Importantly, the results of both models for Amman show an inverse relation between income level and benefits; the income elasticity of net benefits evaluated at the means is about minus 0.4 in both models. On the other hand, there is a pattern of net benefits *increasing* with rent level in the model using the measure of net benefits computed using the household's perception of the market rent for its unit, with the rent elasticity of net benefits being about 1.0. Hence, the relationship between income and net benefits is somewhat ambiguous when the household's perception of market rents is involved. For other urban areas, the income variable is insignificant, but there is a strong inverse relation between rent level and net benefits in one of the two models; the rent elasticity of net benefits evaluated at the means is an enormous minus 5.7.

There is a very significant pattern regarding length of tenure that is

Table 10.4   Regression models explaining net benefits dependent variable: calculated net benefits

| | Amman | | Other urban areas | |
	Perceived[a]	Estimated[b]	Perceived	Estimated
Constant	5.92	5.12	18.31	15.11
Income	−0.0087*	−0.021**	−0.014	−0.0041
	[−0.41][c]		[−0.49][c]	
Time in unit				
3–5 years	3.51	3.19	−4.24	0.268
6–10 years	2.72	7.63*	3.73	−0.861
10+ years	27.48**	30.78**	10.68**	9.64**
Market rent				
(*PmQc*)	0.107**	0.025	0.055	−0.217**
	[1.06]		[5.69]	
Household size				
5–6 persons	−2.04	0.442	−0.960	−0.012
7+ persons	−2.69	−2.074	0.771	0.921
R squared	0.26	0.40	0.13	0.31
R squared (adj)	0.23	0.38	0.05	0.25
F	8.78	16.78	1.74	5.29
(sign.)	(0.000)	(0.000)	(0.11)	(0.000)
d.f.	176	176	83	83

Notes:
a. Net benefits based in part on tenant's perception of market rent.
b. Net benefits based in part on market rent for unit estimated from hedonic functions
c. Elasticity of net benefits evaluated at the mean.
** Significant at 0.05 level or higher.
* Significant at 0.05 to 0.10 level.

evident in all four of the models estimated: no statistically significant rent discount accrues to households who have been in the same unit for less than a decade. Since a substantial majority of renters move more often than this, most households do not obtain any significant benefit from remaining in the unit for several years. Moreover, an examination of cross tabulations of length of time in the unit by various income groups shows that poor renters (incomes less than JD100 per month) in Amman are no more likely than other renters to remain over ten years in a unit. In other urban areas, however, poor renters are about three times as likely as other renters to live in a unit for over a decade.

To elucidate further the kind of patterns implied by the results of these regressions, the net benefits accruing to a half-dozen 'representative' households in Amman and in other urban areas have been computed. For these calculations, various values of the independent variables were substituted into the regression models estimated using the hedonic-index-based estimates of the market rent for the current unit (*PmQc*). For the Amman cases, all variables were fixed except for income level and length of time in the unit. For the cases for other urban areas, only market rent (*PmQc*) and length of time in unit were varied.

The results of this exercise are given in Tables 10.5 and 10.6. For Amman, income was varied by plus and minus JD100 from its mean value of JD272 per month; and we set the length of time in unit at three to five years and over ten years. Table 10.5 shows that the reduction in net benefits associated with higher income is very modest. For households living in their units for three to five years, the increase in net benefits associated with a JD200 decrease in incomes is only JD4.20 (JD6.68–2.48) or about 10 per cent of the average contract rent. In contrast, the difference associated with occupying the same unit for over ten years compared with three to five years is JD27.50 or about 62 per cent of average contract rent.

The results for other urban areas (Table 10.6) display a quite similar pattern but in a somewhat muted form. The increase in net benefits associated with a decrease in rent level from a standard deviation above the mean to a standard deviation below the mean (from JD47 to JD75) is JD6.07 or about 19 per cent of average contract rent. The tenure

Table 10.5 Net benefits to households in Amman: market rent set at mean value and household size set at 5–6 in all cases

	Cases					
	1	2	3	4	5	6
Monthly income (JD)	272	272	172	172	372	372
Years in unit	3–5	10+	3–5	10+	3–5	10+
Net benefits (JD)	4.56	32.06	6.68	34.27	2.48	30.07

Note: Coefficients from models estimated for *NB* defined using *PmQc* estimated with the hedonic models.

Table 10.6 Net benefits to households in other urban areas: income set at mean value and household size set at 5–6 persons in all cases.

	Cases					
	1	2	3	4	5	6
Market rent (JD)	61.3	61.3	47.0	47.0	75.0	75.0
Years in unit	3–5	10+	3–5	10+	3–5	10+
Net benefits (JD)	1.28	10.65	4.39	13.76	−1.68	7.69

Note: Coefficients from models estimated for *NB* defined using *PmQc* estimated with the hedonic models.

effects are again larger. Unit occupancy for over 10 years compared to 3 to 5 years carried with it an increase in benefits of JD9.37 or 29 per cent of average contract rent.

In short, Jordan's system of rent controls has been effective in reducing the rents of long-term occupants, but these benefits have not been equitably distributed. The gains to lower income households, while certainly not trivial, are modest in comparison to those of all long-term occupants (only 20 per cent of long- term occupants are lower income households). Hence, the objective of 'protection of the poor' is being achieved only to a limited degree, and more affluent renter households are enjoying substantial benefits from the existing system of controls.

The potential effects of rent controls on housing supply are often viewed politically as less important than the direct effects on the rents that tenants pay. However, if a rent control regime reduces profits substantially, supply effects may become very important over time. Specifically, if it becomes unprofitable to own rental real estate, landlords may defer maintenance of their existing holdings, and ultimately allow rental units to drop out of the habitable stock. In addition, depressed profits may discourage landlords from investing in new construction, preventing the rental stock from expanding over time.

PROFITABILITY

While it is generally rather difficult to measure the impacts on supply with precision, there are useful data which can be marshalled without too much difficulty. This section summarises some effective methods for examining the effects of rent controls and allied tenant protections on the profitability of rental properties, the overall supply of rental units, and unit maintenance. The objective is to illustrate the different types of information and analytic methods that can be developed for the analysis of individual rent control regimes.

Since the impacts of rent controls on housing supply operate primarily through the effects on landlord profits, a key calculation is the estimate of the rent reduction imposed on landlords from controls. This can be computed from the data derived in the last section as $PmQc$ minus $PcQc$. For Jordan, the rent reductions are on the order of 15 per cent, although the range is wide.[9]

More extensive data were available from a comprehensive analysis of the impacts of rent controls outside the developing world in Washington, DC (Turner, 1990), which was conducted in 1987/88 under contract to the city government as part of a thorough reconsideration of whether rent controls should be continued in their present form, modified, or eliminated. The project gathered data from a wide range of sources. City administrative files provided financial statements for a sample of 814 rental properties, and these were supplemented by the owners or managers of 244 properties who responded to a written questionnaire requesting additional information about property characteristics and finances. In many circumstances, it may not be feasible to gather reliable information on project finances, but, in Washington, the Association of Apartment Building Owners encouraged its members to cooperate because it felt that the financial records would make a compelling case for the elimination of rent controls.

The analysis of DC financial statements began with the estimation of projects' cash flow as a percentage of current equity. The District of Columbia's rent control programme allows landlords to file for a 'hardship rent increase' if their annual cash return falls short of 12 per cent. Thus, it came as something of a surprise that most controlled rental units in the District generated cash returns of less than 12 per cent of current equity. In fact, 22 per cent of all controlled units actually generated negative cash flow. Small properties in particular (fewer than five units) were likely to generate little or no positive cash flow.

However, cash flow is by no means the sole source of return on rental housing investment in the United States. Property appreciation substantially increases the economic return on most categories of controlled rental property in the District, and (at least until the enactment of new federal tax legislation in 1986) the rate of return on investment in rental property was largely tax-free. Therefore, data from the District's tax assessor was used to estimate average rates of property appreciation for rental buildings in different size categories, and the tax treatment of these properties was approximated, using fairly conservative tax accounting principles. Table 10.7 presents the resulting estimates of annual after-tax return on current equity for average rental properties of different sizes in Washington, DC.

These estimates suggest that typical controlled rental properties in the District of Columbia generate quite healthy after-tax return on equity, despite their low cash flow. In fact, estimated profit rates compare favourably to the after-tax returns on competing investment opportunities.

To determine what profits would have been in the absence of controls, the DC study drew upon estimates of likely market rent levels. Market rents were estimated using a three-step procedure:

1 Using data collected prior to the imposition of rent controls, hedonic

Table 10.7 After-tax return on equity: average per unit per year (1985)

Property size	Cash flow($)	Appreciation($)	After-tax return equity(%)
1–2 units	−181	2,441	10.4
3–4 units	110	1,805	18.7
5–9 units	601	1,044	12.4
10–19 units	688	566	11.7
20–49 units	578	567	10.3
50–99 units	1,063	754	11.8
100–249 units	1,153	349	12.4
250 + units	854	−199	10.1

Source: Financial statements for 814 controlled rental properties in the District of Columbia

equations were estimated, expressing rent levels as a function of unit characteristics.

2 The characteristics of units currently available for rent were substituted into the hedonic equation, to estimate the pre-control rent levels for existing units.

3 These pre-control rent levels were inflated to current terms, using the rate of rent inflation typical of uncontrolled central city housing markets in the region, as compiled by the US Bureau of Labour Statistics.

This methodology indicated that the District's rent control programme reduced gross rent levels (rents plus utilities) by $95 to $100 per month on average. Rent reductions of this magnitude translate into significant revenue losses for landlords. Specifically, the study estimated that in the absence of controls, rent revenues to DC landlords would be about 33 per cent higher on average.[10]

The increased rent revenues that would prevail in the absence of controls would ultimately increase property values as well, so that the impacts of controls on the returns to investment in rental property are more complex than they may at first appear. The analysts assumed that the ratio of rent revenues to assessed property values that prevailed in 1987 would have been essentially the same in the absence of controls. Thus, higher revenues translated directly into higher value estimates. It was also assumed that loan to value ratios would be roughly the same, so that both equity and debt would increase if property values were higher. Finally, estimates of appreciation benefits in the absence of controls assumed no change in appreciation rates, but applied the prevailing rates to the estimated values of properties in the absence of controls.

Based on these assumptions, Table 10.8 summarises the estimated impacts of rent control on key financial attributes for average units in different building size categories. After adjusting for changes in equity, interest costs, and property taxes, these estimates indicate that DC housing providers would realise annual increases in net income ranging from about $600 per unit in small properties to about $1350 per unit in large controlled properties. At the same time, annual appreciation gains would probably increase by amounts ranging from just over $100 per unit in large properties to $800 per unit in the smallest properties. If these increased revenues were entirely devoted to raising the investment returns to DC landlords, the profitability of the average large rental property would rise by as much as 5 percentage points, while the profitability of smaller properties would rise by only 1 or 2 percentage points.

TRENDS IN THE NUMBER OF UNITS

In addition to the analysis of profitability, it is worthwhile to look for direct indicators of the effects that rent controls may be having on the supply of rental units. One such indicator is simply the number of units in the rental stock. However, inferring causal linkages between the presence of controls and the size of the rental stock is difficult. On the one hand, controls may discourage investors from expanding the supply of rental units, and may even accelerate the removal of units from the habitable stock. On the other hand, substantially reduced rent levels may increase the number of households who choose to be renters, thereby increasing pressure for more rental housing units. Innumerable other supply and demand factors (interest rates, construction costs, income levels, and demographic trends) all make it difficult to attribute trends in the size of the rental housing stock to the impacts of controls.

In some instances, cross-sectional data can provide clearer insights into what is happening. In Jordan, for example, the survey described earlier documented the continuation of a very high vacancy rate (about 10 per cent nationwide) which has existed at least since 1979. However, the rate among units being offered for sale or for rent was less than half this level. About 60 per cent of those responding to a question about why a vacant unit was not being offered on the market indicated that their units were being held for a family member, often someone working outside of the country. Presumably, if these owners had greater confidence that they could regain the use of their units when they

Table 10.8 Calculating after-tax returns in the absence of rent control for the average controlled unit in buildings with different numbers of units (US$)

	1–2 units		3–4 units		20–49 units		100–249 units	
	Actual	Market	Actual	Market	Actual	Market	Actual	Market
Revenues	4,573	6,082	3,053	4,060	4,179	5,558	4,811	6,399
Value	60,534	80,510	22,564	30,010	12,308	16,370	17,010	22,623
Equity	33,394	44,414	10,981	14,605	10,370	13,792	10,668	14,188
Property taxes	692	920	273	363	201	267	239	318
Interest	1,416	1,883	1,003	1,334	384	511	480	638
Appreciation	2,441	3,246	1,805	2,400	567	754	349	464
Total expends	4,754	5,766	1,805	2,400	3,601	4,676	3,658	5,147
Net income	−181	316	110	348	578	882	1,153	1,252
Income/ equity	−0.54%	0.71%	1.00%	2.39%	5.57%	6.40%	10.81%	8.82%
(Inc.+App.)/ equity	6.77%	8.02%	17.44%	18.82%	11.04%	11.86%	14.08%	12.09%
After-tax return	10.35%	10.97%	18.65%	19.34%	10.25%	10.66%	12.37%	11.38%

wanted to, they would make them available for rent until their family members returned to Jordan.[11] If this is the case, then the reduction in supply is due not to rent controls as such but to the strong tenant protections that often accompany rent control legislation.

As part of the DC analysis, cross-sectional data were combined with data on trends in the number of rental units to gain insights on supply effects in Washington. This study was able to employ data from two points in time: a 1974 general purpose housing survey conducted by the Bureau of the Census and a special survey of 3000 rental units conducted in 1987 as part of the rent control study. In addition, the Washington experience was compared to the experience of several other, unregulated central city housing markets in the United States. During the 1970s, the total number of rental units in the District declined substantially: from 199,100 in 1970 to 170,500 in 1981. This precipitous decline in the size of the District's rental housing stock is often cited as evidence that rent control chokes off supply, despite the fact that new construction is entirely exempt from controls in Washington. However, two key demand-side forces clearly contributed to the decline in the size of the District's rental housing inventory as well. First, the total number of middle- and upper-income renters in the District dropped substantially due to the loss of households to the suburbs and the rising rate of homeownership among those remaining in the District. At the same time, among renters for whom the option of homeownership was unaffordable, incomes failed to keep pace with inflation. Thus, the top end of the rental market was eroded by a decline in the aggregate level of demand, while the bottom end of the market was eroded by a decline in the real purchasing power of low- and moderate-income renters.

Evidence of comparable stock losses in other, uncontrolled central cities over the same period, supports the conclusion that the local regulatory environment is certainly not exclusively responsible for the current shortage of low- and moderate-cost rental housing in the District. The 1970s was a period when the opportunities for both home ownership and suburbanisation were unusually attractive throughout the country. During the 1970s central cities throughout the country lost renter households (and housing units) to the combined attractions of home ownership and suburbanisation.

Demand for rental housing in Washington began to stabilise during the first half of the 1980s. However, the supply of rental units continued its decline. The result was a dramatic drop in the rental vacancy rate from 6.2 per cent in 1981 to 2.5 per cent in 1985. More recently, the District's rental stock started to grow in size. An inventory of all rental units added to or lost from the District's rental stock was conducted, using city records for the 1985–87 period. This inventory revealed that the rental stock was experiencing a net increase of about 800 units annually. Thus, the resurgence of demand for rental housing, reflected in falling vacancy rates, appears to have attracted new investment in the District's rental housing market, resulting in a lagged market response to renewed levels of effective demand.

It is possible that this response might have occurred more quickly in an unregulated rental market, but the responsiveness of District investors to the changing demands for rental housing, and the similarity of recent trends in uncontrolled central city markets, suggests that rent control is not a determining factor in investment decision-making, although it certainly may be a consideration. Recent investors in new and substantially rehabilitated rental units reported (in telephone interviews) that they did not see rent control as a deterrent, and only about one-quarter of those interviewed advocated the elimination of rent control. More cited the District's system of eviction protections, which is seen to reflect a strong 'pro-tenant' bias in the District's regulatory environment, as a serious cause for concern. In addition, the cost and availability of land for new construction and/or structures for substantial rehabilitation, as well as the availability of long-term financing, were cited by the owners of new DC rental units as constraints on their ability to build or substantially renovate housing in the District.

MAINTENANCE

The potential impacts of rent controls on dwelling maintenance can be explored in several ways. The basic hypothesis is that the longer a unit has been occupied by a given renter, the deeper the reductions in rents and, hence, the smaller the incentive for the owner to provide maintenance. If survey data including information on the condition of dwelling units are available, then the condition of units occupied for different periods can be contrasted. One needs to be careful in making such comparisons in multi-unit buildings, however, since it may be the average length of tenure for all the units in the property that will determine some types of maintenance. Upkeep of individual units, including painting and the like, can, however, be measured on a unit-by-unit basis. In addition, if available, administrative records on tenant complaints about maintenance problems can be examined. Again, longer sitting tenants are expected to file complaints with a higher frequency.

In the rent control study conducted for the District of Columbia, surveys of renter households and housing conditions, conducted in 1974 and 1987, made it possible to observe trends in the condition of the rental stock. In addition, both landlords and tenants were asked about the impacts of rent control on maintenance expenditures and maintenance quality. This evidence produced conflicting signals regarding the impacts of rent control on the maintenance of rental housing in the District. On the one hand, DC landlords indicated that the average per unit reduction in rent revenues attributable to controls would be sufficient to correct deferred maintenance problems for roughly half of the DC units that are currently in less than adequate physical condition. Moreover, among owners who reported that their units are in deteriorated physical condition, or that current levels of maintenance expenditures were not sufficient to preserve the quality of their properties, insufficient revenue was the most frequent explanation.

On the other hand, there was no objective indication that a decade and a half of local rent control had resulted in the deterioration of DC rental housing. Since 1974, the overall share of physically deficient rental units actually declined; from 26 per cent to 20 per cent. Moreover, DC units that were exempt from controls exhibited a higher rate of deficiencies than those subject to controls (25 per cent versus 20 per cent), despite the fact that uncontrolled units are typically newer and more expensive. Finally, tenants did not complain that landlords are neglecting housing maintenance. In fact, many indicated that the existing system of controls makes them more confident about asking landlords to correct physical deficiencies in their units.

The foregoing indicates the difficulty of giving any simple response to the question, 'What are the effects of rent controls on urban housing markets?' Glib responses from either politicians or economists are likely to be inaccurate. At the same time, it has been shown that, with some ingenuity and effort, it often is possible to address many of the most important questions raised about the impacts of controls. With concrete information in hand, policy-makers can decide whether the benefits from controls (which may be modest or poorly targeted in some cases) outweigh the costs (which have also proved to be quite modest in many cases).

ACKNOWLEDGEMENT

The Editors wish to record their thanks to the University of Wisconsin Press for permission to use material which has appeared in modified form in Struyk, R. (1987) 'The distribution of tenant benefits from rent control on urban Jordan', *Land Economics*, 64: 125–34.

NOTES

1 For a general review of rent control in developing countries, see Gilbert (1989), ch. 4.
2 Some of the assumptions required to derive this formula are somewhat restrictive. Possibly the most notable is the assumption that the price elasticity of demand for housing services is minus 1. Recent analysis for developing countries suggests a somewhat lower value (Malpezzi and Mayo, 1985). Use of the minus 1 value permits some simplification in the formula to estimate benefits. A comparison of the value of the net benefits (*NB*) computed using the mean values of the right-hand side variables (rather than using each observation to compute net benefits and then taking the mean) with equation [1] and their calculation with a price elasticity of minus 0.7 show differences that range from 1 to 12 per cent.
3 The use of logarithms in the formula comes from approximating the integration of the consumer surplus area under the demand curve.
4 The survey is described fully in Ministry of Planning (1987).
5 Whereas multi-collinearity in hedonic models reduces the precision of the coefficient estimates, the overall predictions remain quite unaffected. See Ozanne and Malpezzi (1985).
6 On the one hand, there may be a wealth effect, in that remittances are being amassed and allow the household to spend a larger share of its current income on housing. On the other hand, the amassing of funds may be making home purchase possible, which might have the effect of reducing current housing consumption as the family tries to increase its savings rate.
7 A potentially serious problem with this calculation is that market or asking rents are increased for the expected length of tenure over which the rents are frozen. While this may be the case, an examination of the rent-to-income ratios did not reveal recent mover households' having to devote extraordinary shares of income to rent in order to secure a unit.
8 At the time of the study, the exchange rate was $US3 = 1 J D .
9 The average net benefit to tenants range from only 43 to 69 per cent of the rent reduction imposed on the landlords.
10 No systematic variations in revenue impacts were found for landlords of different types of properties of different sizes.
11 For more on the high vacancy rates in Jordan, see Struyk (1988).

11 Contingent valuation

Estimating the willingness to pay for housing services: a case study of water supply in southern Haiti

Dale Whittington, John Briscoe, Xinming Mu and William Barron

INTRODUCTION

For most households, access to a high quality, reliable source of water is one of the most important components of the bundle of goods which constitute improved housing. However, progress in improving the quality and quantity of water used by people in rural areas of the developing world has been unsatisfactory in two respects: (1) supplies which have been built are frequently neither used correctly nor properly maintained and (2) extension of improved service to unserved populations has been slow. Though this poor record is not the result of a single factor, a major impediment to improved performance is inadequate information on the response of consumers to new service options. The behavioural assumptions that typically underlie most rural water supply planning efforts are simple. It is commonly assumed that so long as financial requirements do not exceed 5 per cent of income, rural consumers will choose to abandon their existing water supply in favour of the 'improved' system. Several reviews by the World Bank, bilateral donors, and water supply agencies in developing countries have shown, however, that this simple model of behavioural response to improved water supplies has usually proved incorrect (Saunders and Warford, 1977; Churchill *et al.*, 1987). In rural areas many of those 'served' by new systems have chosen to continue with their traditional water use practices.

If rural water projects are to be both sustainable and replicable, an improved planning methodology is required which includes a procedure for eliciting information on the value placed on different levels of service; and tariffs must be designed so that at least operation and maintenance costs (and preferably capital costs) can be recovered. A key concept in such an improved planning methodology is that of 'willingness to pay'. If people are willing to pay the full costs of a particular service, then it is a clear indication that the service is valued (and therefore will most likely be used and maintained) and that it will be possible to generate the funds required to sustain and even replicate the project. Most attempts to incorporate willingness to pay considerations into project design have, however, been *ad hoc*, in large part because of the absence of validated, field-tested methodologies for assessing willingness to pay for water or any housing or service good in the context of rural communities in developing countries.

Two basic theoretical approaches are available for making reliable estimates of households' willingness to pay, but neither has been adequately tested in the field. The first, 'indirect' approach, uses data on observed water use behaviour (such as quantities used, travel times to collection points, perceptions of water quality) to assess the response of consumers to different characteristics of an improved water system. Several modelling approaches are possible candidates here, among them varying parameter demand (Vaughn and Russell, 1982), hedonic property value (Freeman, 1979), and hedonic travel cost models (Cicchetti *et al.*, 1972; Deyak and Smith, 1978). The second, 'direct' approach, is simply to ask an individual how much he or she would be willing to pay for the improved water service, for instance, a public standpost or yard tap. This survey approach is termed the 'contingent valuation method' because the interviewer poses questions within the context of a hypothetical market.

The contingent valuation method can be used to estimate households' demand for improved housing services other than water supply (for example, improved sanitation). One of its advantages is that it can be used in situations in which a good or service is not currently available to a particular group of households. In such cases, the indirect approach is not feasible because there are no market data from which can be derived estimates of changes in the consumers' welfare from the introduction of the good. Conventional wisdom has been that contingent valuation surveys are unreliable because of

> the pervasive feeling that interrogated responses by individuals to hypothetical propositions must be, at best, inferior to 'hard' market data, or, at worst, off-the-cuff attitudinal indications which might be expected to reflect efforts by individuals to manipulate the survey to their selfish ends (Cummings *et al.*, 1986).

In the specific case of rural water supplies, the World Bank concluded more than a decade ago that 'the questionnaire approach to estimating individuals' willingness to pay has been shown to be virtually useless' (Saunders and Warford, 1977). There was, however, little empirical evidence to support this conclusion. Our research objective was to see if contingent valuation surveys could, in fact, be used in developing countries to develop useful estimates of willingness to pay for water services.

A village in southern Haiti was the field site of the study. After describing the specific research area in Haiti, existing water supplies and the water use practices of the population are summarised. The research design and field procedures are then described. The results of the analysis of the sample population's contingent valuation bids lead to general conclusions and remarks on policy implications of the research.

THE STUDY AREA

In August 1986 the research team conducted a contingent valuation survey and source observations in Laurent, a village in southern Haiti. Southern Haiti was selected as the research area because the United States Agency for International Development was funding a rural water supply project there. Moreover, decisions about the level of service and the choice of technology for water supply systems are particularly difficult in Haiti. Per capita incomes there are the lowest in the Western Hemisphere; in 1980 more than two-thirds of the population of five million people had per capita annual incomes less than US$155. Most individuals simply cannot afford the costs associated with private connections. Haiti thus provides a field setting similar to the situation in much of Africa and some parts of Asia, and conditions where an accurate understanding of the willingness of the population to pay for rural water services is likely to be particularly important for sound investment decisions. The Haiti project was designed to provide services to about 160,000 individuals in 40 towns and villages. The project was executed by CARE, which as the implementing agency was responsible for site selection, construction, and community organisation. The CARE project's standard village water supply project was a gravity-fed system supplied with water captured from a mountain spring, feeding a few public standposts in a rural community. CARE provided experienced enumerators for the household surveys, logistic support, and valuable advice on a wide range of issues, from questionnaire design and translation to data on local water use customs. The affiliation of the research effort with the ongoing CARE project provided us with access to villages and justified the study's presence to the local population.

The village of Laurent is located about 15 kilometres from Les Cayes, the provincial capital of southern Haiti. The population of Laurent is about 1500. The region is mountainous, with numerous streams draining into the Caribbean Sea. The rainy seasons are October–January and May–June. The study was conducted during the middle of the July–September dry season.

The population of Laurent consists primarily of small farmers who cultivate sorghum, beans, corn, rice, manioc, sweet potatoes, plantains, yams, coconuts, mangos, and vetiver (a crop used in the production of essential oils for perfume). Few people have regular wage employment, and remittances from relatives and friends living abroad or in Port-au-Prince are common. Eighty per cent of the population of Haiti are illiterate; the illiteracy rate in the study area is probably even higher. Malnourishment is widespread among children in Laurent. The typical family lives in a three-room mud house with plastered walls and a thatched or tin roof.

WATER SOURCES AND WATER USE PATTERNS AND CUSTOMS

Inhabitants of Laurent have access to several sources of fresh water. There are seven sources within approximately 2 kilometres of most of the population: one protected well and six springs in dry river beds.

The springs provide only modest amounts of water, and individuals often wait more than an hour to draw supplies. The average 3 kilometre round trip to a water source can sometimes take several hours.

The population of Laurent expresses strong preferences for clean drinking water and sometimes will walk considerable distances past alternative sources to collect drinking water from sources which are considered pure. Water for drinking and cooking is usually collected by women and children and carried home in relatively standard-size containers (about 20 litres for adults). Although children under five years old are usually bathed at home in basins, adults and older children have a strong preference for bathing in rivers. Clothes washing is usually done in rivers. Some individuals actually pay for public transport to make the roughly 10 kilometre round trip to the nearest river in order to do laundry.

RESEARCH DESIGN

The research design was developed to test whether contingent valuation surveys could be used to estimate water demand relationships suggested by consumer demand theory, and thus reliably to estimate individuals' willingness to pay for improved water services. Economic theory suggests that an individual's demand for a good is a function of the price of the good, prices of substitute and complementary goods, the individual's income, and the individual's tastes, usually measured by the individual's socio-economic characteristics. In CARE's water supply project, the characteristics of the good – public standposts or private connections – are the same for everyone. There is no volumetric charge for water from public standposts; an individual can use as much water as desired. Whether or not a household demands water from the public water system thus depends on the price charged for access to the new system or for participation in the project. If the charge is higher than a given household's maximum willingness to pay (WTP), the household will elect not to use the new water system. Maximum willingness to pay will vary from household to household and should be a function of all of the variables in the demand function except the price of the good itself. The households' WTP bids should thus be positively related to income, the cost of obtaining water from existing sources, and the education of household members; and negatively correlated with the individual's perception of the quality of water at the traditional source used before the construction of the improved water supply system. It could be hypothesised that the WTP bids of women respondents would be higher than those of men because women carry most of the water; for example, a survey in Zimbabwe showed that women were willing to pay 40 per cent more than men for an improved water supply (Ministry of Energy and Water Resources Develpoment, 1985); but alternative interpretations are certainly possible.

The research design attempted to test whether WTP bids are systematically related to the variables suggested by economic theory. If the variation in bids cannot be explained by such variables, three logical explanations can be offered. First, economic theory may not be an appropriate conceptual framework for explaining the behaviour and preferences involved. Second, economic theory may be correct, but the contingent valuation method may not be a sound method for collecting information to estimate the water demand relationships suggested by such theory. Third, errors in execution of the research, such as poor questionnaire design, could lead to invalid inferences about the relationship between the WTP bids and the independent variables.

Because contingent valuation surveys have seldom been attempted in developing countries, the research design was constructed to test for the existence and magnitude of several types of threats to the validity of the survey results. The major problem with the contingent valuation method is that for a variety of reasons, respondents may not answer willingness to pay questions accurately and thus not reveal their 'true' willingness to pay (Cummings *et al.*, 1986). The question format itself may affect the bids. In a pre-test of the questionnaire, different ways of asking the willingness to pay questions were tried. Open-ended, direct questions were tried: for example, 'What is the maximum you would be willing to pay per month to have a public standpost near your house?' In addition, two forms of bidding games in which a series of yes–no questions were asked; for example, 'Would you be willing to pay $X per month for a public standpost near your house?' The appendix presents an example of the sequence of questions used in one of the these bidding games.

Attempts were also made to test for the existence and magnitude of three types of biases in contingent valuation surveys which have been of particular concern in the literature: strategic bias, starting point bias, and hypothetical bias.

STRATEGIC BIAS

Strategic bias may arise when an individual thinks he may influence an investment or policy decision by not answering the interviewer's questions truthfully. Such strategic behaviour may influence an individual's answers in either of two ways. Suppose the individual is asked how much he would be willing to pay to have a public standpost near his house.

If he thinks the water agency or donor will provide the service if the responses of individuals in the village are positive, but that someone else will ultimately pay for the service, he will have an incentive to overstate his actual willingness to pay. On the other hand, if he believes the water agency has already made the decision to install public standposts in the village, and the purpose of the survey is for the water agency to determine the amount people will pay for the service in order to assess charges, the individual will have an incentive to understate his true willingness to pay.

Most attempts to estimate strategic bias have been highly structured experiments in which one group of respondents is told one set of factors about a situation which minimises their incentive for strategic behaviour, and another group receives a different set which maximises their incentives for strategic behaviour (Bohm, 1972). In fact most of the available evidence from the United States and Western Europe fails to support the hypothesis that individuals will act strategically in answering contingent valuation questions, but there is no evidence with respect to developing countries.

Because the surveys were conducted within the context of CARE's ongoing rural water supply project, it was impossible to construct a counterfactual situation (it would have entailed deceiving the study population about CARE policies). Instead, an attempt was made to estimate the magnitude of strategic bias in the following way. The study population was divided into two groups: one group was read the following statement which was intended to minimise strategic bias:

Opening Statement A: I am going to ask you some questions in order to know if you or someone from your household would be willing to pay money to ensure that the CARE Potable Water Project will be successful in Laurent. We would like you to answer these questions at ease. There are no wrong answers.

The water system is going to be managed by a committee of people from Laurent. This committee will be chosen by the people of Laurent. CARE has decided to help Laurent by constructing a water system in this community. Your answers cannot change the fact that CARE has decided to build this water system. CARE never demands money from those people who collect water from public fountains. You will not have to pay money at the public fountains. We need you to tell the truth in order for CARE to construct the best water system for Laurent.

The second group was read another statement which was accurate but left more questions about the purpose of the study unanswered:

Opening Statement B: I am going to ask you some questions in order to know if you or someone from your household would be willing to pay money so that the CARE Potable Water Project will be successful in Laurent. The water system is going to be managed by a committee of people from Laurent. This committee will be chosen by the people of Laurent. The committee will decide the amount each household will have to pay to operate and maintain the water system.

The hypothesis was that if individuals acted strategically, then bids from those who received the second statement would be lower than bids from those who received the first, because the former would fear that a high bid would result in a higher charge by the community water committee.

This is not in fact a strong test for strategic bias because the differences in the two statements are quite subtle. In an ongoing field test in Brazil, strategic bias is being tested for by comparing WTP bids from two different villages, one in which the water authority has already promised to construct a new water system and another in which the water authority has not yet determined whether to build a new system. A comparison of WTP bids from these two villages should be a more conclusive test for strategic bias, assuming that it is possible to control for other differences between the two villages. In future research, it would be useful to have follow-up interviews with selected respondents to see whether the differences which the questions were intended to suggest were understood. An in-depth anthropological research effort might also elicit information on what types of strategic 'thoughts' passed through the respondents' minds during the interview process. If strategic behaviour is found to exist, anthropological research might also yield insights into how to minimise it during the interview.

STARTING POINT BIAS

In the bidding-game question format, the interviewer starts the questioning at an initial price. A respondent who is unsure of an appropriate answer and wants to please the interviewer may interpret this initial price as a clue as to the 'correct' bid. Starting point bias exists if this initial price affects the individual's final willingness to pay. To test for starting point bias we distributed three different versions of our questionnaire, each with different initial prices in the bidding game. The questionnaires were randomly distributed in the sample population.

HYPOTHETICAL BIAS

Hypothetical bias may arise from two kinds of reasons (Cummings *et al.*, 1986). First, the respondent may not understand or correctly perceive the characteristics of the good being described by the interviewer. This has been a particular problem when the contingent valuation method has been used to measure individuals' willingness to pay for changes in environmental quality because it may be difficult for people to perceive what a change, for example, in sulphur dioxide or dissolved oxygen means in terms of air or water quality. This source of hypothetical bias is not likely, however, to be significant for most public services in developing countries. Many rural water systems have already been built in southern Haiti; respondents in the study were all familiar with public water fountains and private water connections and readily understood the possibility that their community would receive a new water system. Moreover, each respondent was shown two colour photographs of public standposts CARE had built in nearby villages. Household members usually studied these with great interest.

Second, it is often alleged, particularly in the context of developing countries, that individuals will not take contingent valuation questions seriously and will simply respond by giving whatever answer first comes to mind. Where this type of hypothetical bias is prevalent, bids will presumably be randomly distributed and not systematically related to household characteristics and other factors suggested by economic theory. The test for hypothetical bias was thus the same as the test for the applicability of consumer demand theory: were bids systematically related to the variables suggested by economic theory?

FIELD PROCEDURES

Fieldwork in the village consisted of two parts: household surveys and source observations. Eight CARE health education promoters and two local college students were trained for two days to carry out the household interviews. Prior to field-testing the questionnaire, a 'focus group' was held in which individuals from a nearby community discussed community water use practices and attitudes. Particular attention was paid in the focus group to household decision-making on water-related matters and to community expectations about operation and maintenance costs. The focus group was not intended, however, to substitute for a pre-test of the questionnaire. The Creole questionnaire was pre-tested extensively in a nearby village before the CARE enumerators were trained, and another day of pre-testing was carried out by CARE staff after training. Because microcomputers were available, revisions to the questionnaire could be incorporated literally overnight, and new copies made for fieldwork the next day.

The majority of households in Laurent were interviewed. Enumerators were instructed to try to interview someone in every house. If no one was at home, a follow-up visit was usually arranged.

The household interview consisted of four sections. The first dealt with basic occupational and demographic data for the family members and summary information on where the family obtained its water. The second section consisted of additional questions on the location of each water source which the family used, perceptions of the water quality at each source, the average number of times each family member went to each source per day, and the number of containers they carried home (the enumerator asked to see the containers used to carry water and estimated their volume). In the third section of the questionnaire, the enumerator read one of the statements used to test for strategic bias and showed the respondent photographs of public standposts CARE had built in other villages. The respondent was then asked for a WTP bid per month (a) for public standposts (assuming no private connections) and (b) a private connection (assuming public standposts were already installed). The fourth section was a series of questions on the health and education of family members and the household's assets (such as whether the household had a radio or a kerosene lamp). The principal investigators and the enumerators had agreed that it was not possible to obtain accurate information on household income through interviews (in fact, the enumerators simply refused to ask either income or expenditure questions because of the antagonism such questions aroused). As a substitute, the enumerator recorded a series of observations about the construction of the house itself, such as whether the house was painted, whether the roof was straw or tin, and whether the floor of the house was dirt or cement.

A detailed map of the village was prepared which indicated the location of all houses and major structures, as well as all water sources. Enumerators who could read maps were given a copy of the village map and asked to assign a number to each household interviewed and to record that number on the map. The enumerator also gave each respondent a ribbon and an index card with the corresponding household number on it and asked the respondent to wear the ribbon or bring the index card on a designated day to the water source used. Enumerators who could not read a map were given a set of ribbons with pre-assigned numbers, dropped at specific points in the villages, and instructed to interview households in clearly specified areas and assign a number to each household interviewed; one of the senior members of the research team then recorded on the map which households were located in the specified areas. Data from household interviews were generally entered into the microcomputer on the same day the interviews were conducted, and processed with dBase III programs. Summary statistics were continually compiled during the course of the fieldwork, and discrepancies in the data and problems with the survey implementation could quickly be detected.

The second part of the fieldwork consisted of observing the quantities of water collected by individuals at all the sources used by the population of the village. The objective of these observations was to verify the information individuals provided in household interviews on the sources they used and the quantities of water collected. Local residents were hired to serve as source observers; they were typically secondary school students on summer vacation. All source observers received one day of training in estimating the volumes of various containers and in recording data in their notebooks. Each time an individual arrived at a source, the source observer recorded household number, name, gender, relative age (adult or child), time of arrival, quantity of water carried away, and whether the individual bathed or did laundry. All sources were observed on the same day from sunrise to sunset. Two shifts of source observers were used for each source. The source observers were monitored closely by the principal investigators to ensure the quality of the data collected.

ANALYSIS OF THE SOURCE OBSERVATION DATA

The analysis of the source-observation data for Laurent increased confidence in the quality of the water-use data obtained from the household interviews. In Laurent data were recorded on 119 trips to water sources by individuals (or groups of individuals from the same household) who identified themselves to the source observers either by wearing a ribbon or displaying an index card with their household number. A comparison was made between the sources these individuals said they used for drinking and cooking in the interview with the source they actually went to on the day of the source observations. Out of the 119 observations, the interview responses were consistent with the source observations for 101 households (85 per cent). In the econometric analysis of the contingent valuation bids, the water source selection data from the household interviews was used to calculate the distance of the household to its primary source of drinking and cooking water.

ANALYSIS OF THE CONTINGENT VALUATION BIDS

In Laurent, 170 questionnaires were completed out of approximately 225 households in the village. The impression gained from sitting in on many of the household interviews is that respondents took the contingent valuation questions, and indeed the entire interview, quite seriously. Fourteen per cent of the households gave an answer of 'I don't know' in response to the willingness to pay question for public standposts; there was a 25 per cent non-response rate for the willingness to pay question for private connections. The mean of the bids in Laurent for the public standposts, 5.7 gourdes per month (US$1.14; US$1.00 = 5 gourdes), seemed realistic. Wildly unrealistic or 'protest' bids did not appear to be present. Based on the pre-test, it was felt that the bidding game question format worked better than the direct, open-ended questions. People generally felt more comfortable with the bidding games and, in fact, the enumerators remarked that the bidding game format was very familiar and easily understood because it was similar to the ordinary kind of bargaining that goes on in local markets of rural Haiti. Hence, in Laurent, only the bidding-game question format was used.

Table 11.1 presents the results of the tests for strategic bias for the willingness to pay questions both for public standposts and private connections. The 150 total responses for public standposts were

Table 11.1 Test for strategic bias

	Opening statement A	Opening statement B
Willingness to pay for public standposts:		
Total observations	77	73
Mean WTP bid*	6.0	5.4
Standard deviation	3.8	3.9
Overall mean	5.7	
Standard deviation	3.8	
t-statistic	1.1	
Willingness to pay for private connections:		
Total observations	67	65
Mean WTP bid	7.5	6.7
Standard deviation	9.0	9.8
Overall mean	7.1	
Standard deviation	9.4	
t-statistic	.5	

Note: Null hypothesis that the two samples are from the same population cannot be rejected at any acceptable confidence level.
  * Mean WTP bid in gourdes per month (5 gourdes = US$1).

Table 11.2 Test for starting point bias

		Starting point	
	2 gourdes*	5 gourdes	7 gourdes
Willingness to pay for public standposts:			
Number of observations	56	47	47
Mean WTP bid*	5.4	6.0	5.7
Standard deviation	3.8	3.9	3.9
$F = 0.32$			
Probability = 0.73			

		Starting point	
	5 gourdes	10 gourdes	15 gourdes
Willingness to pay for private connections:			
Number of observations	48	41	43
Mean WTP bid*	6.7	7.4	7.1
Standard deviation	8.3	8.8	11.0
$F = 0.06$			
Probability = 0.94			

Note: Null hypothesis that the three samples are from the same population cannot be rejected at any acceptable confidence level.
* In gourdes per month.

relatively evenly divided between statement A (77 responses) and statement B (73 responses), as were those for the private connections. As anticipated, for respondents who received statement A, the mean bids both for public standposts and private connections were higher than for those who received statement B, but the difference is not statistically significant. On the basis of this test, the hypothesis that respondents were not acting strategically when they answered the willingness to pay questions cannot be rejected.

Table 11.2 presents the results of a similar statistical test for starting point bias. If starting point bias were a problem, it would be expected that the low starting point (2 gourdes for public standposts; 5 gourdes for private connections) would result in a lower bid, and that the high starting point (7 gourdes for public standposts; 15 gourdes for private connections) would result in higher bids. The mean bids in Table 11.2 do not appear to vary systematically with the starting point. The null hypothesis that the three samples are from the same population (that there is no difference in the responses from individuals who received different starting points) cannot be rejected, although the confidence intervals are wide.

On the basis of these results, there was no reason to attempt to adjust the WTP bids for strategic or starting point bias. The mean of WTP bids for the public standposts was 5.7 gourdes per household per month. Assuming an average annual household income in Laurent of 4000 gourdes (US$800), the mean bid is about 1.7 per cent of household income and is significantly lower than the 5 per cent rule of thumb often used in rural water supply planning for maximum 'ability to pay' for public standposts. The mean of WTP bids for private connections, 7.1 gourdes, was not much higher (2.1 per cent of household income), but these bids are based on the assumption that the public standposts are already in place.

PROBIT MODEL OF CONTINGENT VALUATION BIDS

Variations in the bids for public standposts and private connections were modelled as a function of the variables thought to affect willingness to pay. To measure income, an ordinal measure of the value of household assets was developed, based on eight questions and observations about the quality of housing construction and household possessions (WLTH), and supplemented by two other indicators of income: (1) whether the household received remittances from relatives living abroad and (2) the occupations of the principal members of the household. In the model, remittance data were simply treated as a dummy variable (FINC). Occupation data were used to group households into two categories (farmers and non-farmers) and were also represented by a dummy variable (IOCP). Education was measured as the sum of the years of school of up to two adults in the household (HHED). From the village map, the distance of each household to its drinking water source was measured (DIST); these distances served as a measure of the cost of obtaining water from the existing source, which was viewed in the model as the 'price' of the close substitute of the improved water service. The measure of water quality (QULT) was based on the respondent's answers to seven questions concerning taste, odour, healthfulness, reliability, colour, dirt, and conflict (quarrels) at the source.

Although the value households place on the proposed water system is a continuous variable, probably the most reliable data generated from the bidding game are the set of yes/no responses to questions about specific, discrete prices. Thus, the observed dependent variable obtained from the bidding game procedure is not the maximum amount the household would be willing to pay but, rather, an interval within which the 'true' willingness to pay falls. Linear regression is not an appropriate procedure for dealing with such an ordinal dependent variable because the assumptions regarding the specification of the error term in the linear model will be violated (Maddala, 1983). Therefore, an ordered probit model, discussed below, was used to explain the variations in WTP bids.

Let V_h be the maximum willingness to pay of household 'h' for the proposed water system. Based on consumer demand theory, V_h can be hypothesised as a function of the attributes of the new and existing water sources and the household's socio-economic characteristics

$$V_h = a + X_h B + e_h, \tag{1}$$

where X_h is a vector of the household's characteristics and the attributes of the sources, a and B are parameters of the model, and e_h is a random term with a standard normal distribution. Since V_h is not observable from the bidding game, equation (1) cannot be estimated. However, from the interview responses the ranges within which V_h will fall are known. Let $R1,..., R_m$ be the 'm' prices which divide the range of WTP space into $m + 1$ categories, and let y_h be a categorical variable such that

$$y_h = \begin{cases} 1 & \text{if } V_h < R_1, \\ 2 & \text{if } R_1 < V_h < R_2, \\ M + 1 & \text{if } V_h > R_m \end{cases} \tag{2}$$

Let $i = 1,...., M + 1$. From equation (1), we have $y_h = 1$ if

$$R_{i-1} < a + X_hB + e_h < R_i \tag{3}$$
or $$R_{i-1} - a < X_hB + e_h < R_i - a \tag{4}$$
or $$(R_{i-1} - a - X_hB)/sd < e_h/sd < (R_i - a - X_hB)/sd, \tag{5}$$

where sd is the standard deviation of e_h. Assuming e_h follows a standard normal distribution, then

$$
\begin{aligned}
P(y_h = i) &= P(R_{i-1} < V_h < R_i) \\
&= P(u_{i-1} - X_hB < e_h < u_{i-1} - X_hB) \\
&= F(u_i - X_hB) - F(u_{i-1} - X_hB),
\end{aligned} \tag{6}
$$

where $u_i = R_i - a$ and $F(.)$ is the cumulative standard normal density function. (Equation (6) is the ordered probit model used to explain the variations in WTP bids.) The maximum likelihood estimates of u_i and B are consistent (Maddala, 1983).

The results of the estimations are presented in Tables 11.3 and 11.4. The chi-square statistics illustrate that the overall models are highly significant. The adjusted likelihood ratio $(1 - \{[L(B) - K]/L(0)\})$ is 0.142 for the model of bids for public standposts and 0.177 for the model of

Table 11.3 Willingness to pay bids for public standposts

	Coefficient	t ratio
Dependent variable: Probability that a household's willingness to pay for a public standpost falls within a specified interval		
Independent variables:		
Intercept	.841	1.350
Household wealth index (WLTH)	.126	2.939
Household with foreign income (FINC = 1 if yes)	.064	.232
Occupation index (IOCP = 1 if farmer)	−.209	−.848
Household education level (HHED)	.157	2.113
Distance from existing source (DIST)	.001	5.716
Quality index of existing source (QULT = 1 if satisfactory)	−.072	−2.163
Sex of respondent (male = 1)	−.104	−5.41
Log-likelihood	−206.01	
Restricted log-likelihood	−231.95	
Chi-square (df. = 7)	51.878	
Adjusted likelihood ratio	.142	
Degrees of freedom	137	

Table 11.4   Willingness to pay bids for private connections

	Coefficient	t ratio
Dependent variable: Probability that a household's willingness to pay for a private connection falls within a specified interval		
Independent variables:		
Intercept	−0.896	−1.344
Household wealth index (WLTH)	0.217	4.166
Household with foreign income (FINC = 1 if yes)	0.046	0.194
Occupation index (IOCP = 1 if farmer)	−0.597	−2.541
Household education level (HHED)	0.090	1.818
Distance from existing source (DIST)	0.000	1.949
Quality index of existing source (QULT = 1 if satisfactory)	−0.099	−2.526
Sex of respondent (male=1)	−0.045	−0.207
Log-likelihood	−173.56	
Restricted log-likelihood	−202.48	
Chi-square (df. = 7)	51.831	
Adjusted likelihood ratio	0.177	
Degrees of freedom	120	

bids for private connections, where K is the number of independent variables in the model. The coefficients for all the independent variables are in the direction expected (e.g. households with higher education or wealth tend to bid higher). The t statistics indicate that the variables for household wealth, household education, distance of the household from the existing water source, and water quality are all significant at the 0.05 level in both models. The sex of the respondent was statistically significant in the model for public standposts, but not in the model for private connections. The results clearly indicate that the WTP bids are not random numbers but are systematically related to the variables suggested by economic theory.

POLICY APPLICATIONS

This ordered probit model can be used to predict the number of households in a community which will use a new source if various prices were charged. Since the interval for each category is known, y_h (the

Table 11.5 Demand schedules for new water sources, derived from the ordered probit models (price versus number of users)

	Price (gourdes per month)			
	2	5	7	10
Number of users:				
Public standposts				
(N = 145)	138	97	68	40
	Price (gourdes per month)			
	5	10	15	20
Number of users:				
Private connections				
(N = 127)	78	62	40	17

category into which household h falls) may be predicted from inequality (4) by calculating X_hB. Summing the number of households in each category in Laurent yields the demand schedules presented in Table 11.5. Such demand schedules are precisely the kind of information needed by planners and engineers to make sound investment decision, and this ordered probit model, estimated with WTP bids obtained from a contingent valuation survey, is a promising approach to modelling village water demand relationships.

SUMMARY AND CONCLUSIONS

The results of this study suggest that it is possible to do a contingent valuation survey among a very poor, illiterate population and obtain reasonable, consistent answers. There does not appear to be a major problem with either starting point or hypothetical bias. The evidence with regard to strategic bias is less conclusive, but neither the admittedly limited test for strategic bias nor the experience of the enumerators indicated that it was a problem.

From this research it is not possible, of course, to judge or prove whether individuals in the villages would in fact pay the amounts they indicated in the contingent valuation survey if a water agency actually tried to collect the money. To do so would require a contingent valuation survey in a village before a water system is built, and then a resurvey after the system was completed and collection efforts made, to compare the prior bids with actual behaviour.

Nevertheless, the preliminary results of this research strongly suggest that contingent valuation surveys are a feasible method for estimating individuals' willingness to pay for improved water services in rural Haiti. This has important policy implications for rural water supply projects such as CARE's because it seems to show that going into a village and conducting a relatively simple household survey can yield reliable information on the population's willingness to pay for improved water services. The implications of these preliminary research findings are not, however, limited to the rural water sector. This research suggests that contingent valuation surveys may prove to be a viable method of collecting information on individuals' willingness to pay for a wide range of public infrastructure projects and public services in developing countries.

APPENDIX: EXAMPLE OF A BIDDING GAME

Here are pictures of CARE public fountains set up in Rosier and Port-a-Piment.

(a) Do you think your household would be willing to pay 5 gourdes each month to use a public fountain located in your neighbourhood?

Yes	Go to (b)
No	Go to (c)
I don't know	Go to (f)

(b) We do not know how much the water committee will decide for each household to pay for using the public fountain each month. If the decision is for each household to give 10 gourdes each month, would your household be willing to pay this?

Yes	Go to (f)
No	Go to (d)
I don't know	Go to (f)

(c) We do not know how much the water committee will decide for each household to pay for using the public fountain each month. If the decision is for each household to give 0.50 gourdes each month, would your household be willing to pay this?

Yes	Go to (e)
No	Go to (f)
I don't know	Go to (f)

(d) Would your household be willing to pay 7 gourdes each month to use a public fountain located in your neighbourhood?

Yes	Go to (f)
No	Go to (f)
I don't know	Go to (f)

(e) Would your household be willing to pay 2 gourdes each month to use a public fountain located in your neighbourhood?

Yes Go to (f)

No Go to (f)

I don't know Go to (f)

(f) Think for a moment, what is the largest amount of money your household would be willing to pay each month to use a public fountain? If it would cost your household more than this amount, your household could not afford to pay and would not be able to use the public fountain.

Amount of money: . . . .

I don't know: . . . .

ACKNOWLEDGEMENTS

This research was carried out as part of the Water and Sanitation for Health (WASH) project, sponsored by the Office of Health, Bureau for Science and Technology, US Agency for International Development. A version of this chapter was presented at the annual meeting of the North American Regional Science Association, November, 1986, in Columbus, Ohio. It has also been published in modified form as 'Estimating the Willingness to pay for water services in developing countries: a case study of the use of contingent valuation surveys in Southern Haiti', (1990) *Economic Development and Cultural Change*, 38, 2, 293–311. The editors gratefully acknowledge the consent of the University of Chicago Press to use the material here.

12 Discounted cash flow analysis
Present value models of housing programmes and policies

Stephen Malpezzi

INTRODUCTION

Housing investment is the largest single form of fixed capital investment in most economies, developing or developed. In developing countries, the shelter sector usually ranges from 10 to 30 per cent of household expenditure, or 6 to 20 per cent of GNP. Housing investment typically comprises from 10 to 50 per cent of gross fixed capital formation. Further, housing investment's share of GDP rises as economies develop.[1]

It is important to be able to analyse housing market and policy issues in terms of a rigorous analytical framework, and, in particular, to compare the efficacy of alternative government interventions in the housing market according to a common and comprehensive set of criteria. This chapter presents such a framework, and illustrates its use with some examples from Malaysia and Ghana. More details on the applications can be found in Malpezzi (1989) and Malpezzi, Tipple and Willis (1990).

Present value analysis can be used to study both owner-occupied and rental housing, public or private. This kind of analysis can compare the costs and benefits of different housing programmes, and, within programmes, facilitates comparisons of programme components. Present value analysis can yield insights on the efficiency of investments, as well as some useful information on equity.

Conventionally, housing is more often regarded as a social service or basic need than as a productive investment. But shelter and infrastructure investments are directly productive: they are investment in an asset which yields a flow of services over time. Efficient investments are those which yield the most services for the resources society puts into them, discounted for when they are available. As is well known, the present value investment rule yields the most efficient set of investments.[2] It is for this reason that present value computations are the heart of the evaluation below. If the present value of the benefits to society (net of taxes, subsidies and other distortions) of a house exceeds the present value of its costs, investment in such a unit is, by definition, economically efficient. More precisely, the returns to such an efficient investment exceed the opportunity cost of capital, measured by the discount rate.

Social equity is related to the distribution of assets and the services from those assets. Purely distributional issues are usually thought of as an essentially political decision, not an economic one. But equity and efficiency are, in fact, more closely related than is often realised. The present value model[3] cannot set society's distributional goals, but it can demonstrate the cost of reaching alternative goals. Properly augmented with information about demand, it can predict whether current policies, or alternatives, are likely to reach distributional goals; i.e. to predict who will benefit and by how much.

In previous papers, the author and colleagues have emphasised the need for careful survey research.[4] Present value models can make use of information from such surveys[5], but they do not require such surveys. This does not mean that the household survey or other approaches are being eschewed, but these models have some obvious advantages. They are cheap and fast. They permit comparisons of many different interventions between quite different units, tenures, and programmes. They can be used to integrate the results of many other kinds of studies, including survey work. They function as a concrete tool that focuses analysts' attention on constraints rather than 'needs assessments'. But perhaps their most important function is to help analysts in governments and donors, to think like developers, but from the point of view of other actors (house purchasers, landlords and tenants; society as a whole) as well as developers. By studying constraints and incentives (and their effect on costs, profitability, and affordability), and redesigning policies accordingly, we can move on beyond projects to the market-wide changes which are required today.

The main ideas behind this model can be expressed quite simply. Many government interventions affect the price of housing: regulations, taxes, and subsidies are the most common interventions. Each intervention has costs and benefits which are not often explicitly estimated. Even when they are costed out, one or two interventions are usually studied in isolation. As the analysis below demonstrates, costs and benefits of different interventions accrue to different participants in the market (developers, homeowners, landlords, tenants), and one individual's subsidy will not always cancel another's cost. This model can help clarify the structure of incentives in a given market and regulatory framework, and predict the results of changes in those incentives.

The next section reviews cash flow modelling and present value analysis, and presents an example of a representative housing investment. The following section presents an example from Malaysia, comparing several representative investments in current programmes, to give the reader a better feel for how to use the model and draw policy conclusions. The penultimate section presents a somewhat different example from Ghana. The final section discusses directions for further work.

PROGRAMME ANALYSIS USING PRESENT VALUES

The present value model is really nothing more than an organising framework which keeps track of many interventions for several investments at once, and of how the costs and benefits of these interventions accrue to various participants in the market (developers, landlords, tenants, home owners) and their implications for the economy as a whole. Because it is basically an accounting framework, the principle of 'garbage in–garbage out' applies. The model is no better than its inputs, although it permits sensitivity analysis to study the effects of changes in assumptions on outcomes.

It should be emphasised from the outset that these models are not 'black boxes,' i.e. users need to understand their inner workings. Fortunately, the heart of each specific model is a conceptually simple cash flow model. An overarching goal of the model can be expressed as follows: to examine government programmes the way private developers analyse their investments.[6]

The intention is to model investments realistically, but there are some approximations in all models. For example, in all models in this chapter, the development period is treated as a single, year long period; but more complex quarterly, or monthly time period models can be developed. Some specific limitations common to all three models include the following:

1 Many interventions, subsidies, taxes, and market imperfections must be costed out beforehand; these are inputs to the models.
2 Risk and uncertainty are not built into the models, but can be studied by varying assumptions parametrically.
3 While the models are somewhat comprehensive (especially Malaysia), each component is necessarily simplified. For example, in their current form these models only handle fixed rate mortgages.

Of course, none of these limitations is immutable. Further development of the models could address any of these. The final section of the chapter will discuss limitations and possible extensions in more detail.

HOW TO MAKE SHELTER INVESTMENT DECISIONS: THE CONCEPT OF PRESENT VALUE

How should society decide how much investment in housing should be made, how much in clothing factories? The present value investment rule yields the most efficient set of investments.

Present values are computed by adding a stream of net costs and benefits from an investment after discounting them to account for the fact that a dollar (or other currency) today is worth more than a dollar tomorrow. Consider a simple four-period example:

$$PV = A_0 + A_1/(1 + r) + A_2/(1 + r)^2 + A_3/(1 + r)^3$$

where A represents the net costs and benefits in each of four periods, and r is the discount rate or the opportunity cost of capital. Consider the following simple example: if an initial investment of $300 is followed by three years of net returns of $150 per annum, and the discount rate is 10 per cent, the present value is:

$$PV = -300 + 150/(1 + 0.1) + 150/(1 + 0.1)^2 + 150/(1 + 0.1)^3 = 73$$

The present value rule states that if the present value of the investment is greater than zero, the investment yields greater than the opportunity cost of capital (the normal rate of profit for an investment of that type), and the investment should be undertaken. Note that the rule applies both to market and planned economies: the only difference is that in a functioning market the present value rule will be followed automatically as investors seek to maximise their returns.

The model based on this rule is conceptually quite simple. First, analyse the notional cash flows from the investment in efficiency (market) prices. The present value of the costs (at market prices) is the real cost of the investment, and the present value of the benefits (again, at market prices) is the market value of the unit. If the market value exceeds the cost, the unit should be built (according to the criterion of efficiency).

SUBSIDIES, TAXES, REGULATIONS AND OTHER INTERVENTIONS

Now, government interventions can be introduced into the model. Government subsidises, regulates, taxes and otherwise intervenes in housing markets for a variety of purposes. After the 'economic' cost-benefit analysis just described (at undistorted competitive market prices), each policy intervention can be analysed in turn by examining how it changes the prices and corresponding present values. Present values have the advantage of enabling direct comparisons of the costs and benefits of quite different interventions in different programmes. Some interventions impose costs (e.g. land use regulations, taxes, rent controls, building regulations) and some benefits (e.g. land subsidies, tax relief, financial subsidies). Some interventions confer corresponding costs and benefits on different market participants; for example, rent controls benefit some tenants at the expense of landlords (and perhaps some other tenants). Other interventions confer costs and/or benefits on some participants without an obvious corresponding gain or loss elsewhere. For example, some very high infrastructure standards can confer large costs on developers without producing much in the way of benefit for anybody.

In this framework, there are three entities from whose point of view housing policies and programmes are evaluated: the economy, housing suppliers (or developers) and households. The exact incidence of the various costs and benefits of government interventions can be a subtle issue. For example, although the incidence of the property tax appears straightforward (property owners pay the property tax) some portion of the tax could be shifted to tenants (for rental property) or to the owners of capital generally (if capital markets were well integrated). Incidence can depend on the competitiveness of the market, the state of transactions costs and knowledge in the market, the efficiency of financial markets in a country, and the time frame; in other words it is rarely settled and unambiguous. In this model a simple approach is adopted, where the entire cost or benefit is assigned to one participant (as will be described for each specific intervention below). If our knowledge of actual incidence improves, it would not be difficult to build in more sophisticated treatment of incidence.

THE HEART OF THE ANALYSIS: THE CASH FLOW MODEL

We begin with two very simple models: one for a rental investment (Tables 12.1 and 12.2) and one for a sales unit (Table 12.3, described later). Table 12.1 is divided into two parts, input data, and the cash flow model itself.

SIMPLE RENTAL CASH FLOW MODEL

Discount Rate	10.0%
COSTS	
Land Price/sq m	$20.00
Lot Size	1600
Construction Cost/sq.m	$15.00
Unit Size	700
Net Depreciation (%K)	5.0%
Maintenance (%K)	3.0%
PRICES	
Market Rent	$1,200
Change in Real Rents	0.0%
Change in Land Price	5.0%
Change in Structure Price	0.0%
INTERVENTIONS	
Land Subsidy/Tax	$6,000
Rental Income Tax Rate	0%
DEMAND	
Rent/Income Ratio	20%

	Period 0	Period 1	Period 2	Period 3	Present Value
ECONOMIC COST-BENEFIT					
Land Cost	-32000				-32000
Structure	-10500				-10500
Maintenance		-315	-315	-315	-783
Market Rents		1200	1200	1200	2984
Salvage Value				45969	34537
Economic C-B	-42500	885	885	46854	-5762
INTERVENTIONS					
Land Sub/Tax	6000				6000
Income Tax		0	0	0	0
Financial C-B	-36500	885	885	46854	238
DEMAND					
Affordable @ Income of	6000	6000	6000	6000	

Table 12.1 Simple rental cash flow model

Table of spreadsheet cell contents (rotated layout):

	A	B	D	E	F	G	H	I
1	|				'SIMPLE RENTAL CASH FLOW MODEL|			'^Present
2								^Value
3	'|Discount	0.1						
4	'COSTS							
5	'Land price	20		'^Period 0	'^Period 1	'^Period 2	'^Period 3	
6	'Lot size	1600	'Land Cost	+-B5*B6				E6+@NPV(+$B$3,F6..H6)
7	'Const cost	15	'Stucture	+-B7*B8				E7+@NPV(+$B$3,F7..H7)
8	'Unit size	700	'Maint Cst		+$E$7*$B$10	+$E$7*$B$10	+$E$7*$B$10	E8+@NPV(+$B$3,F8..H8)
9	'Depr (%K)	0.05						
10	'Maint (%K)	0.03	'^Mkt Rent		+$B$12*(1+$B$13)^F4	+$B$12*(1+$B$13)^G4	+$B$12*(1+$B$13)^H4	E10+@NPV(+$B$3,F10..H10)
11	'PRICES		'Salvage				+-E6*(1+B14)^H4-(E7-(H4*B9*E7))*(1+B15)^H4	E11+@NPV(+$B$3,F11..H11)
12	'Mkt Rent	1200						
13	'DP, rents	0	'Econ C-B	@SUM(E6..E11)	@SUM(F6..F11)	@SUM(G6..G11)	@SUM(H6..H11)	E13+@NPV(+$B$3,F13..H13)
14	'DP, land	0.5						
15	'DP, struct	0	'Land Sub	+B17				E15+@NPV(+$B$3,F15..H15)
16	'INTERVENTIONS		'^Tax		+-F10*$B$18	+-G10*$B$18	+-H10*$B$18	E16+@NPV(+$B$3,F16..H16)
17	'Land Subs.	6000						
18	'Tax Rate	0	'Fin C-B	@SUM(E13,E15..E16)	@SUM(F13,F15..F16)	@SUM(G13,G15..G16)	@SUM(H13,H15..H16)	E18+@NPV(0.1,F18..H18)
19	'DEMAND							
20	'Rent/Income	0.2	'Afford. @ Income of		+F10/$B$20	+G10/$B$20	+H10/$B$20	

Table 12.2 Rental cash flow model, contents of spreadsheet cells

The inputs are:

1 Land price per square foot, and lot size, in square feet;
2 Construction cost per square foot, and unit size, in square feet;
3 Depreciation (net of maintenance), and maintenance costs, both as a percentage of structure cost;
4 Market rent per annum;
5 The annual percentage change in rents, land prices, and structure prices.[7]
6 Two government interventions: an up-front land subsidy, and a tax on rental income. All other interventions, including other taxes, and finance, are ignored for now.

The cash flow model has four columns representing four periods: a development period, and three periods during which the unit is rented out. At the end of the period, the unit is sold for a salvage value which is determined by the original cost of the unit, the real rates of price increase for land and structures, and the depreciation of the structure. For simplicity, we assume the developer is the same as the landlord, although it would not be hard to change this assumption. All values are real.

The rows of the cash flow model start with the pure 'economic' costs and benefits of the housing investment, followed by a series of government interventions. These can comprise, for example, land subsidies, construction subsidies, the full set of regulatory costs, finance, and taxes. Each intervention changes the cash flow. In this model, for simplicity, we have only two: a land regulation, equivalent to a tax, and taxes on rental income. Finally, there is a simple 'demand side.' Given the periodic rents, and an assumed rent-to-income ratio, it is straightforward to compute the income at which a particular unit is 'affordable.'

Key results from this simple model are as follows. From the point of view of the economy as a whole, costs of this particular investment are greater than the benefits (5762 units of account greater), so the unit is an inefficient use of society's resources.[8] But the net effect of government intervention (a large land subsidy, zero taxes) is to provide developers with a net incentive to build such inefficient units. The (over) simple demand calculation predicts that households with incomes of about 6000 will live in such units.

Of course there are many ways to extend such a simple model: more periods, more interventions, a better demand side; rather than go on at length, some extensions are illustrated in the next sections. But note that even this extremely simple model can yield useful information as one step up from 'back of the envelope' calculations.

SIMPLE SALES CASH FLOW MODEL

		Period	Period	Period	Period	Present
		0	1	2	3	Value
Discount Rate	10.0%					
COSTS		ECONOMIC COST-BENEFIT				
Land Price	$10.00	LandCost	-16000			-16000
Lot Size	1600	Structure	-10500			-10500
Construction Cost	$15.00	Market Value		30000		27273
Unit Size	700					
PRICES		Econ C-B	-26500	30000	0	0	773
Market Value	$30,000	DEVELOPER COST-BENEFIT				
Sale Price	$25,000	LandTax/Sub	-5000			-5000
INTERVENTION						
Land Subsidy/Tax	($5,000)	Developer C-B	-31500	25000	0	0	-8773
FINANCING		PURCHASER COST BENEFIT				
Loan Term	3	Downpayment	-2500			
Loan to Sales Price	90.0%	Finance		-8262	-8262	-8262	-22500
Interest Rate	5.0%					
Market Rate	5.0%	Purchaser C-B	19238	-8262	-8262	-8262	4453
DEMAND						
Debt Service to Income	20%	Affordable @ Income of:	41311	41311	41311	
Down Payment to Income	100%	GOVERNMENT COST	5000			

Table 12.3 Sales cash flow model

While the model described in Table 12.1 can be worked out with pencil and paper, it is convenient to build such a model with a spreadsheet such as *Lotus 1-2-3*, *Quattro*, or *Supercalc*. For larger models (discussed below), using a spreadsheet becomes almost essential. Table 12.2 is the same as Table 12.1, except that the spreadsheet cell entries rather than their results are presented. Except for the labels the cells in the cash flow table contain formulae, and the input area values. More details about the computer implementation of such models can be found in Malpezzi (1989). Readers can input Table 12.2 in their own spreadsheets for practice. Note that once the model is built, only data in column 'B' change. Readers can then explore the implications of changing costs, taxes, and the other inputs for efficiency, the developer's decision, and tenant income.

A simple sales cash flow model is presented in Table 12.3. Note that finance has been added (of course landlord finance could easily be added to the previous model). Most costs and benefits are all captured in the first period (construction) and the second period (sales), except that benefits from finance accrue over the life of the mortgage. In other words, all benefits from future (imputed) rents, future maintenance costs, and future taxes are capitalised into the market value in the sales period.

Here the two market participants are developers and home purchasers. Given the particular assumptions for this investment, the benefits exceed the costs to the overall economy, by about 773 units of account. But the developer pays a land tax, and the sales price is regulated, so that the financial cost-benefit is negative; developers will not build such units, even though they are efficient.

While these two models are quite 'small' all other models are only variants of them; more time periods can be added, more interventions, more complicated financing, a more realistic demand side, graphic presentation of results, and so on. As the models get larger, the spreadsheet implementation gets more complex, but the underlying ideas are the same.

ANALYSIS OF A MALAYSIAN HOUSING PROGRAMME

In 1988, the government of Malaysia requested the World Bank to undertake a study of why formal housing costs appeared to be so high. This was despite the existence of the Special Low Cost Housing Programme (SLCHP) designed to induce private developers to build low cost housing. The programme was innovative in that virtually all the units were to be built by private developers on both private and state-owned land. Developers were permitted to build 60 per cent of their units to high standards with high profit margins in order to subsidise the remaining 40 per cent. For the low cost units, reduced infrastructure standards and streamlined regulatory approvals were envisaged but these proved to be insufficient. In its first year, the SLCHP fell well short of the target number of low cost housing units it had hoped to build.

The resultant study had three interlocking parts: an analysis of aggregate market and macroeconomic data; an analysis of the costs, benefits and incidence of a wide range of government interventions; and a detailed study of land use regulations. The analysis of government interventions' effects was carried out on 14 representative housing units (inside and outside the programme, formal and informal, owner occupied and rental, and in different locations) and demonstrated that the costs of regulatory and pricing restrictions far outweighed the benefits of subsidies and regulatory exemptions.

Important market conditions and programmatic features have changed since the original study, so specific results here do not necessarily represent current conditions, but examining the model remains instructive. The model starts with the standard economic cost/benefit analysis of a representative investment, and then adds the major interventions with simple assumptions about incidence:

The Economy

> \+ Market value of the unit
> \– Resource cost to the economy
> _____
>
> Net economic cost-benefit

The Developer

> \– Resource cost to the economy
> \+ Land subsidy
> \+ Development period infrastructure subsidy
> \+ Construction subsidies (materials, finance, etc.)
> \– Cost of land use and building regulations (including delays)
> \– Land acquisition, other taxes
> \+ Sales price
> _____
>
> Net financial cost-benefit to developer

House Purchasers

- − Sales price
- − Registration taxes
- − Property taxes
- − Extra transactions cost of programme participation
- + Market value of the unit
- + Recurrent infrastructure subsidies
- + End user finance subsidies

Net financial cost-benefit to purchaser

The actual cash flow model is just an extension of Table 12.3; it is too large to reproduce in full here.[9] Once again, the relationship between these calculations and market behaviour is straightforward. If the economic cost/benefit is positive, the unit is efficient. If the developer's cost/benefit is positive, a supply response will be observed. If the purchaser's cost/benefit is positive, there will be demand for the units.

An exemption from a regulation which has an identifiable benefit to society similar to its cost is treated as a subsidy. Reductions in regulations which do not yield corresponding benefits are, therefore, pure cost reductions. In other words, there is a baseline of 'normal' desirable regulation from which extra regulatory costs are measured. Defining this baseline is necessarily subjective. A conservative approach is to focus attention on regulations which clearly have negligible benefits (as measured by market values or demonstrable externalities).

PRESENT VALUE ANALYSIS OF THE SPECIAL LOW COST HOUSING PROGRAMME (SLCHP)

The present value analysis demonstrated why developers found the Malaysian SLCHP less than enticing. Even more importantly, the analysis illuminated why costs were so high for formal units outside the programme.

Some SLCHP developments received land below market cost from individual states; others received land at closer to opportunity cost or used private land. For this particular representative example, it is assumed that state land was used for a nominal charge. The difference between fees charged and estimated market land value yielded a land subsidy of about M$8000 to the developer (M$2.5 = US$1 approximately). Charges for infrastructure connections were about M$1450 below their estimated cost. A small cement subsidy roughly cancelled cement prices above world market prices, and no other significant construction subsidies were found.

However, the developer's land and infrastructure subsidies were largely offset by other interventions. The costs and benefits of land use regulations were estimated in a separate exercise using a well-known land use model, the Bertaud model (see Bertaud, *et al.*, 1988). That model compares the cost of development under current regulations with the cost of development under some regulatory baseline, adjusted for any benefits which may accrue from the additional regulation.

In Malaysia land use and infrastructure standards are particularly high. The SLCHP produced a new set of lower standards for low cost housing. However, the actual approval of plans utilising the new standards is given by local authorities who have so far continued to rely on previous, higher standards. The analysis examined the costs and benefits of the following interventions:

1   Reduction in saleable land from numerous requirements for road widths, setbacks and large set-asides for public areas. The estimate of the cost of land use and infrastructure regulation is M$6000 per plot which is the difference in cost between the current standard (as little as 25 per cent saleable land) and the recommended standard (65 per cent). This is a conservative estimate. For example, by lowering road design standards, saleable land is increased, but surfacing and maintenance costs are also reduced. The latter cost saving is not included.

2   The delay imposed by regulatory procedures which tie up capital and increase risk. Developers often take several years to receive planning permission. Given estimates of the average delay (compared to some reasonable baseline), the amount of capital tied up, and its opportunity cost, estimating this cost is straightforward. In the example, a very conservative estimate of M$1000 is used.

3   Controls on sale prices are a regulatory cost to the developer but a financial subsidy to the purchaser. The nature and size of the transfer depends on location, since sales prices vary less than market values with location. In this particular example the unit is worth about M$30,000 but the sales price is M$25,000, implying a transfer of M$5000.

4   Other costs to developers include building codes and standards (judged to be not large in the present case, since these codes seem roughly in line with the market) and regulations encouraging sale to ethnic Malays and indigenous peoples which are costly to comply with, especially in some urban areas with high Chinese and Indian populations. Compliance lengthens the developer's holding period, and frequently discounts have to be offered to reach the desired mix.

These regulations are estimated to add M$1625 to developers' average unit costs.

5 Taxation of housing in Malaysia is fairly light. The main taxes for sales programmes comprise acquisition taxes, assumed to be borne by the developer, and property taxes, assumed to be borne by the purchaser. Capital gains taxes are levied on nominal, not real, appreciation on a sliding scale depending on the holding period. In this example it is assumed the unit is not sold so no tax is paid. Despite the light tax environment, more work on taxation should be high on the agenda for future model development.

6 Financial subsidies are estimated by calculating the present value of the subsidised cash flow at unsubsidised rates. In some countries with poorly developed financial markets or particularly inappropriate terms (for example, 30-year fixed rate instruments) a baseline unsubsidised rate may not exist; a range of estimated market rates can then be tested. Fortunately in Malaysia's well-developed financial system reasonable comparators existed, making the calculation of the financial subsidy straightforward.

Financing is analysed on the basis of fixed rate self-amortising mortgages. For this example, a 25-year loan is assumed, a market rate of 12 per cent (a conservative assumption; the market rate for such a loan could be as high as 14 per cent), a nominal lending rate of 10 per cent and a loan to value ratio of 0.95. To calculate the present value of the financial subsidy, deflate the nominal principal and interest payment in real terms, then take the present value of the initial loan followed by the real repayment stream, discounted by the market rate of interest.

THE INCIDENCE OF INCENTIVES

Market valuation of units is straightforward in Malaysia, where there is an active resale market; economic cost was arrived at by subtracting net regulatory costs from financial cost. In the absence of any government interventions, it is estimated that the unit would cost about M$28,100 but be worth about M$30,000.

Figure 12.1 summarises the net incentives and 'disincentives' faced by developers of a representative SLCHP unit. The developer receives substantial subsidies through low cost land and reduced infrastructure standards, but these are more than outweighed by the costs of regulations and the pricing restriction that effectively requires the unit to be sold below cost. The net effect of these interventions is to add about M$4000 to the developer's cost (the bottom bar). Given the above

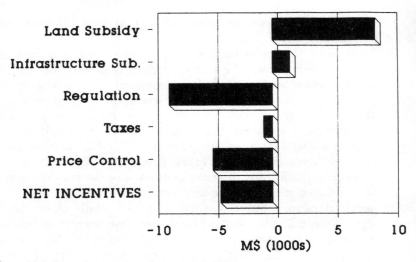

Figure 12.1 Incentives to the developer: SLCHP unit, Selangor, current standards

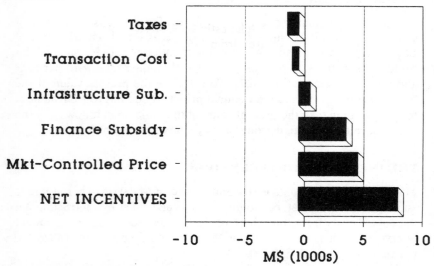

Figure 12.2 Incentives to the purchaser: SLCHP unit, Selangor, current standards

result, that in the absence of interventions the unit would be worth about M$2000 more than its cost, the extra M$4000 leads to a net loss on each unit of about M$2000.

Figure 12.2 tallies the incentives and disincentives to purchasers of a

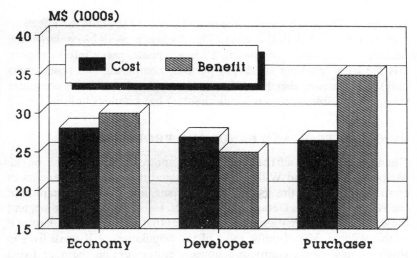

Figure 12.3 Present value model results: SLCHP unit, Selangor, current standards

representative SLCHP unit. The estimated subsidy to the purchaser of nearly M$9000 mostly derives from below market pricing restrictions and mortgage financing.

Figure 12.3 shows how these add up from the point of view of the economy, the developer and the purchaser. This particular unit is efficient, in other words, it benefits the economy more than it costs it. Demand would be strong in the absence of additional purchaser incentives, but would be very high given the additional subsidies involved. But because of regulation, developers lose money, so they would build these efficient units only if forced to do so (for example, to obtain planning permission for other units) or if purchasers paid higher than official prices. In Malaysia, the former predominates but both mechanisms can be studied with such models.

Analysis of units outside the programme was equally instructive in demonstrating how regulations hamstrung developers and ultimately consumers. Analysis by location revealed that administrative pricing led to excess demand in some areas, and costly inventories of unsold units in others.

A PRESENT VALUE MODEL OF CONTROLLED RENTAL HOUSING IN GHANA

The previous section presented a model which analysed the costs and benefits to individual developers and house purchasers. In another exercise, a different present value model was used to study the effects of rent control in Kumasi, Ghana. The Malaysia model was large (30 periods, two tenures, comparing 15 investments, many interventions); this model is 'medium sized' (10 periods, comparing two rental units only). This model also has a better demand side, and the computer version (available; see note 9, this chapter) has graphics built in.

RENT CONTROL AND LANDLORD PROFITABILITY

Ghana has had one of the strictest rent control regimes in the world. Malpezzi, Tipple and Willis (1990) evaluated the costs and benefits of controls in Kumasi, the second city of Ghana, and traditional capital of the Asante kingdom. Details of the market, the rent control regime, and the cost-benefit estimation can be found in that paper and in Chapter 8 of this volume. Very briefly, most of the population of Kumasi lives in single rooms within compound houses, either renting them or living with kin. Rents for this ubiquitous accommodation have been controlled since the Second World War; since 1986 the rents have been set at 300 cedis per month (C90 = US$1 at 1986 official rates). Despite strict controls, and serious problems in land, finance, materials and other regulations such as inappropriate building codes, some rooms have been added to existing compounds in recent years though few new houses have been built.

Malpezzi, Tipple and Willis (1990) used hedonic estimation on a small 'uncontrolled' part of the market[10] to estimate what units would rent for in the absence of controls (roughly 600 cedis for the typical unit in 1986, the date of the survey data). Simple demand models were used to estimate the rents households would pay in the absence of controls (roughly 1100 cedis for the typical household, implying that typical households would consume more housing if it were available at market prices). From these, standard consumers surplus measures were constructed of the static costs to landlords and the benefits to tenants (typical cost to landlords of roughly 300 cedis, but much smaller benefit to tenants, near zero under some assumptions, because the subsidy is offset by reduced consumption).

The market effects of rent control can be analysed using a simple cash flow model of a representative rental investment. Such an approach yields information on the rate of return to such investments under different control regimes and market conditions. The estimates from the static cost-benefit analysis can be used as inputs to the model; and since they are only point estimates from data several years old, they can be varied to study changes since that time, and for other kinds of units.

The model is described more completely in Chapter 6 of Malpezzi Tipple and Willis (1990). Landlord–developers are assumed to build or purchase a unit in development period (year 0) and rent out the rooms therein for ten years. During this time landlords collect rents and spend money on maintenance and taxes. At the end of the ten year period the unit (structure and land) have some salvage value.

This model is quite simple, yet it allows two different rent regimes to be compared (referred to below as controlled and uncontrolled, but two different control regimes can also be compared). The interaction between controls, taxes, maintenance, depreciation, profitability and affordability can also be examined in a simple but consistent framework.

Any number of alternatives for relaxing or removing controls may be specified by changing the time path of rents in the revised case. Here the model assumes a simple but dramatic quadrupling of rents to 1200 cedis per room, which then rise with inflation. Other assumptions for this par-ticular example include a general inflation rate of 20 per cent per annum; a real discount rate of 10 per cent; and land and structure values rising with inflation (i.e. their relative price remains constant). Wages, depressed for so long, are assumed to rise by 2 per cent per annum. Gross depreciation of the unit is assumed to be 4 per cent per annum; spending more on maintenance is assumed to reduce net depreciation one for one. Landlords pay an income tax of 5 per cent on rents collected, rising to 10 per cent for higher rents. Households at the median income are assumed to be willing to spend 8 per cent of their income on such a unit; the income elasticity, 0.6, assumes that lower income households spend higher fractions, and vice versa.[11]

GAINS AND LOSSES FROM FOUR COMPONENTS AND THEIR INTERACTION

Five key components of landlord cash flow are: initial outlay, rents, maintenance, taxes, and capital gains. Initial outlay does not change when controls are removed, but the other four do. Rent control directly reduces profitability because it reduces the rents a unit can command. But reduced rents also affect maintenance (and depreciation), taxes, and capital gains. These 'indirect' effects can be large and should be taken into account. Figure 12.4 (an output from the model) summarises the changes.

Rents

Ghana's current rent control regime fixes nominal and reduces real rents. Under the assumptions presented above, the reduction in real

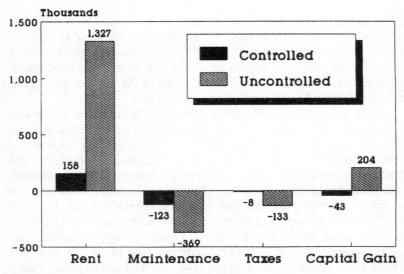

Figure 12.4 Components of landlord's present value: Kumasi

rents increases over time as inflation takes a larger bite. Our 'uncontrolled' or modified regime assumes rents rise with inflation. Figure 12.4 shows that the present value of real rents collected from this 15 room compound over 10 years under the decontrolled regime (real rents at 1200 cedis per month) is about 1.2 million cedis, compared to 160,000 cedis under controls (nominal rents at 300 cedis, 20 per cent inflation).

Taxation

If taxes on rental income are collected from landlords, rent control reduces these taxes as it reduces rent. This partially offsets the reduction in rent to landlords, but also decreases government revenue. Effective property taxes for units of this type (when collected) are on the order of 2000 cedis. It is assumed that this remains unchanged in real terms over time. Figure 12.4 shows that the present value of taxes rises from 8000 cedis to about 125,000 cedis. From the landlord's point of view the tax increase partially offsets the rents collected. But it also represents a badly needed increase in government revenue.

Maintenance

Landlords have the option to increase or decrease maintenance. While good data are lacking, it is assumed in these first simulations that maintenance on a controlled unit is a minimal 1 per cent of structure cost. When controls are removed landlords increase maintenance to a still modest 3 per cent. Figure 12.4 reflects this assumption that landlords spend about three times as much on maintenance if controls are removed. But if the unit is not maintained it depreciates faster. This will reduce the capital gain.

Capital gains

Capital gains (and losses) stem from several sources. First, structures and land can appreciate more or less than general inflation. The simulation presented here assumes both structure and land prices move with inflation. Second, the land and the structure may originally (at period 0) be worth more or less than the value of resources put into it. The assumption here is that the original value of the structure (in the baseline case) is equal to its financial cost to the landlord–developer, but that the land is worth considerably more. The latter can readily be the case given the traditional system of land allocation, especially if the landlord is an Asante local to Kumasi. Thus one incentive is immediately identified in the current system: in traditional areas building a house can give the landlord control over land worth far more than the fees and ground rents paid. Third, the real value of structural capital declines as the unit depreciates. As has been shown, depreciation depends on maintenance, and in this version a simple one-for-one offset is assumed. Fourth, if the rent control regime changes, increases in rents will be capitalised into value.[12] Finally, note that it is not assumed that the landlord actually sells the unit; among other things, we can abstract from capital gains taxation.

EFFECTS OF RENT CONTROL ON LANDLORD'S PROFIT

Figure 12.5 presents the internal rate of return of the rental investment, or discount rate at which the present value of the unit is zero (the landlord–investor breaks even). Under controls, the rate of return is roughly zero; in the absence of controls it is about 8 per cent. In other words, under controls landlords could at least maintain the value of capital with housing if other investments were yielding negative returns. Without controls, housing could compete for capital with investments

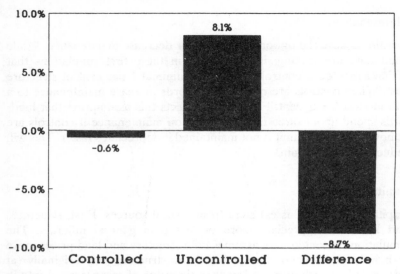

Figure 12.5 Internal rate of return, new rental: Kumasi

yielding up to 8 per cent. Controls reduce the rate of profit by about 9 per cent overall.

ALTERNATIVE INVESTMENT OPPORTUNITIES

The present value model can illustrate the effect controls have on the rate of return on rental units. Often analysts are interested in the effect of controls on supply which depends further on two key variables: alternative investment opportunities, discussed here, and responsiveness to changes in rates of return (about which more later).

Financial investments have consistently shown negative returns in Ghana (see Figure 12.6), averaging minus 40 per cent during the period in the figure. A zero rate of return can be quite attractive in such an unstable environment. Returns to some other activities, such as trading, were probably higher than housing. But not everyone can trade, and if one has a large sum to invest alternatives are required.

Of course real financial rates in Ghana are no longer on the order of minus 100 per cent as they were in the 1970s. Ironically, as inflation abates and financial investments yield positive real returns, the capital gain/ preservation motive for housing investment will weaken. Current cash flow will become a more critical investment incentive in a stable economy. Rent control will bite deeper into incentives for such investment as economic recovery proceeds.

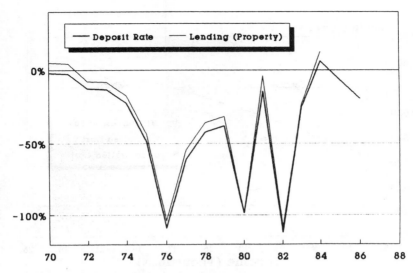

Figure 12.6 Real interest rates, 1970–86: Ghana

AFFORDABILITY

The model has a simple demand side which enables a study of the affordability of each rent regime. Given an income distribution (midpoints of income quintiles) and average and marginal propensities to consume housing (the median income household's average willingness to pay for such a unit, and the income elasticity), the model generates willingness to pay for the entire income distribution and compares it to rents under each regime. Figure 12.7 presents this graphically for the first year. The diagonal line represents willingness to pay by income, given the demand parameters above and income distribution data from the 1986 survey. The five dots on the line represent the income midpoints of the five quintiles. Note that controlled rents (the bottom horizontal line) are affordable to all 5 quintiles, while higher uncontrolled rents are affordable to the top 2 quintiles.

Recall that this is a representative new unit, in 1986 units of account. 'Representative' does not mean that all units will deliver the same package of housing services, or will rent for this amount. Some units will be produced which will rent for more, some for less. Note in particular that rents are always higher for new units in uncontrolled markets. The model was also used to examine other units, including existing units, and to study different decontrol options by varying the time path of rents in the alternative case; for details see Malpezzi, Tipple and Willis (1990).

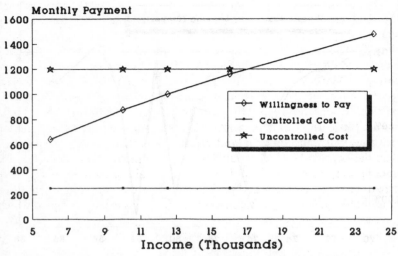

Figure 12.7 Willingness to pay for a new rental unit, first year: Kumasi

Finally, note here, that successful decontrol requires action to improve the supply response of the market. Finance, land, materials, and building and planning codes all need urgent attention.

FUTURE DIRECTIONS

Future directions for World Bank models

All models can be improved. As some of the models developed by the author and his colleagues have been offered as examples for other model builders, it may be worth discussing some of the improvements which could be built into future models. Financial side improvements would include incorporating variable rate mortgages, graduated payment mortgages, and other mortgage instruments, and building in a term structure for discount rates.

These models currently tally up costs and benefits to developers, house purchasers, landlords and tenants. While costs and benefits to government (on and off budget) can be derived from the model output, a more explicit presentation of the budgetary implications of different investments could be built in. These could distinguish on and off budget expenditures, and could differentiate between central and local government.

One of the most important improvements which could be made for the analysis of government programmes is the incorporation of consumer's surplus into the demand side of the model. The basic idea is quite simple. The models presented above measure financial costs and benefits to different actors in the market who may value transfers differently (see Mayo *et al.*, 1986). For example, governments may tie a subsidy to the purchase of a particular kind of housing unit, and target a particular household type. If (as often happens) the target households are not those who would pay the market price for such a unit, we can infer that they will value such 'in kind' transfer less than the market; i.e. less than the household which would bid a market price. In extreme (but not uncommon) cases, we would find that certain programmes are very expensive to government and society, but their corresponding benefit to households may be very small. Other programmes can be designed which increase target household welfare by an equal amount for a lower cost; or alternatively increase particpant welfare much more for an equivalent cost. Adding simple consumer's surplus calculations to the demand side permits explicit calculation and comparison among alternative programmes.

Currently, the model assumes that one discount rate is used to discount all cash flows.[13] Another possible refinement of the model is to introduce different interest rates for different actors (for example, one for households, one for builders and developers, and one for government) and/or different rates depending on risk (i.e. the variance of the expected cash flow).

These are among many possible improvements to the model. A summary list of the above, and other possible improvements, would include the following:

1 construct a lease-with-option-to-buy cash flow model to complement the rental and sales models;
2 make holding periods flexible and endogenous;
3 improve output presentation, and include graphic output;
4 improve the demand side of the model, including building in consumer's surplus measures;
5 permit parametric changes in assumptions about incidence;
6 improve treatment of finance, including graduated payment and variable rate mortgages, to complement the current level payment fixed rate mortgage;
7 provide a richer menu of alternative discount rates for different cash flows, and improve the treatment of risk generally;
8 explicitly total up budget and off-budget costs for different levels of government.

Building your own model

The chapter has, it is hoped, demonstrated the versatility of the present value/cash flow approach to studying housing markets, especially when augmented by a demand side. By now the reader should be aware that while the principles of building these models are straightforward, there are a number of design choices which have to be made to apply the models to given country, market conditions, and purpose of the analysis. It is not recommended that you take any of the existing spreadsheet models described above and plug in the data for other countries; why, when it is so easy to build your own model?

Readers who are potential model builders should study other models as they build their own. Malpezzi (1989) and Malpezzi, Tipple and Willis (1990) also provide more details on the construction of these models. Related readings include, for example Jaffee (1985) (especially good on finance) and Bertaud *et al.* (1988) (the Bertaud land use model). A few simple principles can guide our development:

1   *Examine other models but build your own.* Before beginning, sit down and list the main questions you want the exercise to answer. Tailor models to the country and to the policy issues.
2   *Start simply and build the model up.* Try to get all the important components in first, then add refinements to each one. For example, in our own models we are just beginning to tackle variable rate mortgages. If others will use it, put more time into menus and 'user friendliness'. If you will be presenting results, build in graphics.
3   *Focus more on design than on results, especially at the beginning.* Do not be afraid to begin with poor data; take your best guess, and let the model guide data improvement. Talk to developers, landlords, and other market participants, at least as much as government officials and professional colleagues.

The last point may be less obvious than the other two. Perhaps the greatest benefit of building these models is that as you think of how to build the model and its inner workings, you begin to focus less and less on 'what is the answer? what is the number?' and more and more on 'how does the market work?'

NOTES

1   See Mayo, Malpezzi and Gross (1986) for a broad review of housing markets and policies in developing countries.
2   It is the durability of assets which requires that present values (or the closely related concept, internal rate of return) be used to correctly analyse capital

investment decisions. Incremental capital output ratios (ICORs), another commonly used criterion, do not account for the durability of assets. Further discussion of the present value rule can be found in any finance text (for example, Brealey and Meyers, 1981) or any economics text with a discussion of capital budgeting (for example, Yotopoulos and Nugent, 1976).

3 Sometimes the word 'model' will refer to the generic present value/cash flow model; sometimes it will refer to a particular implementation, for example, the 'Malaysia model'. The difference will be clear from the text.

4 Malpezzi, Bamberger and Mayo (1981).

5 See especially Malpezzi, Tipple and Willis (1990).

6 These models are similar to those actually used by developers, except that developers properly only consider their own cash flow, whereas here a complex set of interactions which accrue to different market participants (developers, landlords, tenants, homeowners) is considered, as well as the pure economic cost benefit of the project.

7 To keep this first model simple changes in the general price level is ignored. Only the relative price of rents, land, and structures, change; i.e. all prices are real. Later models incorporate background inflation as well, because the real prices of some cash flows which are fixed in nominal terms depends on inflation.

8 If an inefficient investment is to be favourably considered because it helps reach some specified equity goal, the equity gain and its associated benefits should be added explicitly to the model. Failing to consider efficiency when in pursuit of equity is a 'cop out', because there are always alternatives with different costs to reach the same equity goal; see the introduction to this chapter.

9 Copies of spreadsheets for these models are available from the author, if you write and send two formatted diskettes. The spreadsheets require Lotus version 2.1 or some compatible spreadsheet.

10 That paper discusses the problems of picking an 'uncontrolled' sample in some detail, and how the 'uncontrolled' rent is affected by controls.

11 See Malpezzi and Mayo (1987).

12 For these simple simulations full capitalisation of the difference between the average net income streams for the two regimes are assumed. Also that decontrol was completely unanticipated. These assumptions could be compared to alternatives for future work.

13 There is one exception. The real discount rate for the present value of mortgage financing is the market nominal rate of interest for such loans, less the rate of inflation.

13 Cost-benefit analysis
Housing and squatter upgrading in East Africa

Gordon Hughes

INTRODUCTION

In making decisions about a specific housing project or policy any government faces the question as to whether the resources required to build/upgrade the housing, subsidise rents or provide better infrastructure and services could be used better for other purposes. The alternatives against which the project must be compared might lie elsewhere in the housing sector or in other sectors – agriculture, transport, health – of the economy. The nature of the alternatives is necessarily uncertain because governments are, of course, not able to identify all possible uses of public resources and then rank them in terms of their return to the economy and to the population in general. Equally, different uses of resources may benefit different sub-groups of the population, so that the choice between projects or policies may in part be a choice between directing the benefits of government action to people who differ in terms of their location, their sources of income, their standard of living and their composition of expenditure. Finally, the nature of the resources required for different projects may differ. Some projects may consume substantial amounts of capital or foreign exchange while others depend upon inputs of skilled labour or other scarce factors.

The role of cost-benefit analysis is to provide a consistent framework which may be used to compare such alternatives in deciding between competing demands on limited investment resources. In essence, cost-benefit analysis has developed as a method of decentralising the decisions which must be made in drawing up a programme of investments for a sector or for the whole economy. We might think in terms of attempting to draw up an optimal investment plan designed to reflect the government's specific priorities as well as the constraints on the availability of various resources. Such an optimal investment plan must be inherently fragile because of the uncertainties under which policies are developed and projects designed. It is, therefore, necessary to provide a basis for evaluating individual projects or policy proposals as they arise, which is consistent with the broad thrust of the government's general investment and development strategy. Modern economists have developed the theory and methods of cost-benefit analysis in this context (see Little and Mirrlees, 1974; Dreze and Stern, 1987; Hammond, 1988) though the origins of cost-benefit analysis lie in the evaluation of specific transport and water projects without reference to a wider planning strategy. The historical background to cost-benefit analysis means that the literature tends to focus upon the evaluation of projects, i.e. specific investment proposals. However, it is important to understand that the same techniques may be used to assess proposals for the reform of policy of all kinds (see Stern 1987). In this chapter, the conventional terminology will be followed by referring to the analysis of housing 'projects' but these may encompass such reforms as the abolition of rent controls, the provision of subsidised housing or mortgage finance, and the modification of tenure arrangements or zoning controls.

The key element in modern cost-benefit analysis which provides the link between the constraints and objectives which determine the overall investment plan and the analysis of a particular project is the use of shadow prices. These emerge naturally from the solution of optimisation problems as indicators of the cost – in terms of the government's objectives – of the constraints on resource availability which are binding at the optimum. Each shadow price tells us, in effect, how much it would be worth paying for an additional unit of the resource; foreign exchange, capital, skilled labour, etc.; with which the shadow price is associated. It is a fundamental result in the theory of economic planning that shadow prices may be used to signal the scarcity of these resources so that production may be efficiently organised if separate agents (individuals, firms, local governments) can be told to maximise the net value of their output taking account of market and shadow prices. The theory of cost-benefit analysis has extended this idea by using shadow prices to allow for both the scarcity of important resources and also the effects of particular distortions in the working of markets which mean that market prices do not appropriately reflect the cost (benefit) to the economy of using (producing) a particular item.

The prevalence of distortionary tax systems and trade regimes in developing countries means that market prices do not provide a reliable indicator of the opportunity costs of the various inputs used by a project or of the social value of the goods and services which will be produced by the project. Further, even in developed countries, imperfections in the working of the labour and similar markets may imply that market wages and factor prices should be adjusted to take account of the secondary effects of the project on the labour or other markets. It is not possible to discuss the computation and application of shadow prices in cost-benefit analysis here, so those interested should refer to manuals such as Pearce and Nash (1981) or Squire and van der Tak (1975) which explain how shadow prices may be calculated. The estimation of a set of shadow prices for the analysis of a housing project is given in Hughes (1976). The crucial point is that, in most cases, all values in the cost-benefit analysis of a project should be estimated using shadow rather than market prices.

There is one distinction between different types of shadow prices which is particularly relevant to housing projects. Some project appraisals use 'economic' or 'efficiency' pricing while others extend shadow prices to encompass what is usually called 'social' pricing. The difference between the two lies primarily in the treatment of the distributional impact of changes in the demand for labour due to the project. Analyses at efficiency prices take no account of the effect of the project on the distribution of income. They are based on the assumption that the government is, or should be, concerned solely to maximise the aggregate level of national income by using resources in the most efficient manner possible. This does not mean that distributional issues should be neglected altogether, but rather an assumption that the government is able to take steps to influence the distribution of income independently of any decisions about such projects. On the other hand, social pricing incorporates a set of distributional weights so that income accruing to low income households because new employment opportunities have been created is given a higher value than income accruing to better-off households. The use of social prices is appropriate if it is not possible to redistribute income by means of taxes or other policies which have little effect on the efficiency with which resources are used in the economy. In the face of such a constraint the government may reasonably wish to take account of the distributional impact of a project when deciding upon the allocation of investment and other resources.

For most sectors the differences between analyses at efficiency and social prices are relatively minor, but the distributional consequences of alternative housing projects may be quite significant. Since some housing projects may be very specifically designed to benefit low income households, estimates of the benefits of such projects may be strongly influenced by the treatment of distributional questions. This will be discussed fully below. It has also been argued that housing projects could be an important source of employment for relatively unskilled workers. This would imply that the costs of housing projects evaluated at efficiency prices might be significantly higher than their cost at social prices. Empirical work suggests that any such differences are unlikely to be large enough to alter the ranking of projects. None the less, one should use social prices to evaluate the costs of housing projects whenever distributional considerations are taken into account on the benefit side so as to ensure a consistent framework for the appraisal.

A source of some confusion for those unfamiliar with the theoretical foundations of shadow pricing is the problem of the 'numeraire'. In any system of pricing it is necessary to choose a reference point or numeraire with all other prices being expressed relative to this basic unit of measurement. In the past the numeraire for market prices was some specified quantity of gold or silver, while now it is a monetary unit defined by law. For shadow prices the choice of a numeraire is more difficult and two are used quite widely:

1 the basket of goods consumed by households in a particular reference group or the average consumption basket for all households evaluated in either case at domestic market prices;
2 convertible foreign exchange expressed in terms of domestic currency at the official exchange rate.

There are various issues which may be considered in choosing between these two numeraires (see Hughes, 1986), but, provided that the set of shadow prices is estimated consistently, the choice of numeraire should have no effect on the ranking of projects. The main difference between the two is that the first generates a shadow exchange rate (SER) for converting foreign exchange costs while the second implicitly defines the SER as being equal to unity and introduces a consumption conversion factor (CCF) which is equal to the reciprocal of the SER under the first system. For convenience, the second numeraire (i.e. foreign exchange) will be adopted in this chapter.

The literature on cost-benefit analysis is very large. Readers interested in obtaining an overview of the methodology or an exposition of the general issues which arise in the application of the technique should refer to a survey such as Papps (1987) or one of the many texts which cover the subject such as Lal (1980), Pearce and Nash (1981), Ray (1984) and Londero (1987). This chapter will focus on the main questions which arise in applying cost-benefit analysis to low income housing projects.

THE COST OF HOUSING PROJECTS

The fundamental elements required for the estimation of the cost of a housing project at either efficiency or social prices are (a) a set of appropriate shadow prices, and (b) detailed financial estimates of the kind prepared by the quantity surveyors or engineering consultants involved in the design of the project. It is usual to present shadow prices in terms of 'conversion factors' each of which is the ratio of the economic/social price for an item to its associated market price. Conversion factors will usually be estimated for groups of goods or services or even for broad categories of project expenditure so that there is a considerable degree of averaging and approximation in estimating the total cost of a project at shadow prices. For a very small number of items, the project analyst may consider that it is worth estimating specific conversion factors because of their importance in the total cost of the project or because of peculiar features of the markets for the items concerned. Apart from such items, the estimation of the costs of housing projects at shadow prices is a relatively straightforward matter provided that sufficient data to carry out a full financial analysis is available.

Any system of conversion factors used for the costs of housing projects will highlight the distortionary effects of domestic taxes and duties on imports or exports. The higher are these taxes, the lower will be the value of the conversion factors for the items concerned. The conversion factors for non-traded goods and services, which will include important building materials such as bricks and concrete blocks, may be low also because of the effect of taxes and other distortions on the prices of the goods and services which are used to produce the items. Substantial differences in the conversion factors for the various elements of building costs may imply that least cost analyses of construction methods and of building specifications may produce rather different results at market and shadow prices. This implies that those designing projects should be careful to base their calculation on relative shadow prices rather than market prices when choosing techniques of production and other aspects of the project specification.

One particular application of this point concerns the labour intensity of production and building methods. Advocates of 'intermediate technology' note that there is substantial under-employment or unemployment in most developing countries. They infer, correctly, that the shadow cost of employing unskilled workers should be much lower than the market wage rate. Thus, it is argued that developing countries should rely upon labour-intensive methods of production or construction wherever possible rather than on modern capital-intensive methods which rely upon machinery and sophisticated building materials. Implicitly, this argument rests on the assumption is that the shadow cost of employing unskilled labour is close to zero, an assumption which is likely to be quite wrong for the urban areas of most developing countries. The reason is that paying additional wages to unskilled workers has a cost in terms of the resources required to provide the goods and services which they will consume out of their wages. When we use social prices, this consumption cost will be offset for the very poorest households by the distributional benefits of improving their welfare. However, most urban households are rather better-off than most rural households, so the distributional benefits of paying extra wages to urban unskilled workers may be quite small. One would also expect that the wage in the informal sector for unskilled workers settles at a level to balance the demand and supply for such labour. An increase in employment opportunities may increase the market wage rate, so that the shadow wage must take account both of the additional consumption of workers directly employed on the project and also that of other workers who receive a higher wage because the wage rate has been bid up.

Labour-intensive techniques may also require more supervision and higher inputs of skilled labour in general. Since the availability of skilled labour is often a major constraint on development, the shadow wage rate for skilled workers tends to be high relative to the market wage rate. These factors may mean that simple rules of thumb, which rely upon simple measures of labour intensity, may be quite misleading when evaluating alternative construction methods or techniques of production. It is essential to carry out a careful costing of the alternatives using a full set of shadow prices. Sometimes, as in Lal's (1978) analysis of rural road-building in the Philippines, one may conclude that labour-intensive techniques do have the advantage, but this is not something that should be taken for granted.

Another common assumption about the costs of low income housing projects concerns the importance of foreign exchange costs. Shortage of foreign exchange may mean that governments have a preference for projects which have a low cost in terms of foreign exchange rather than ones which involve the importation of substantial quantities of machinery, materials or other inputs. In this respect low income housing projects have substantial appeal because most of the direct costs are local, so that their foreign exchange component may appear to be rather low. The role of the shadow exchange rate or its inverse is, of course, to provide a systematic basis for weighting foreign exchange costs against other costs but there is still a tendency to rely upon rather simplistic rules of thumb. These may be quite misleading because indirect foreign exchange costs may be as important as direct ones.

For a typical low income housing project the direct foreign exchange costs will include expenditures on construction machinery, building materials and house fittings which are used by the project. A concern with foreign exchange costs will, thus, tend to favour projects which use local materials and labour-intensive construction techniques on the grounds that they absorb little foreign exchange. Such appearances may be seriously misleading. 'Local' materials and fittings may be manufactured from imported intermediate goods, while labour used in construction may be withdrawn from the production of agricultural goods which could be exported or which would replace imported foodstuffs in urban markets. These indirect effects of a project on the availability of foreign exchange may be complex and can only be measured on the basis of a detailed analysis of shadow prices as illustrated in Hughes (1976). In the case of the Kenyan case study the share of direct and indirect foreign exchange costs in total costs varied from under 25 per cent for sewers and roads to over 40 per cent for water supply. The variation was significant even for different types of housing and, excluding traditional mud houses, there was little correlation between the share of foreign exchange costs and cost per square metre for different types of housing. Thus, it would be wrong to assume that low income housing will tend to absorb less foreign exchange per $1000 of total costs than housing for middle or higher income households.

THE THEORY OF HOUSING BENEFITS

The starting point for any evaluation of the benefits of a housing project must be an estimate of the value to households of the housing and other services which will be made available as a result of the project as measured by their willingness to pay for these services. At its simplest this means that we will be concerned with the total revenue from rents that will be earned by the agency responsible for letting the housing which has been built for the project. The issue is usually more complicated than this because most housing projects consist of a number of separate elements which may affect the housing market in different ways.

Some of the difficulties involved in assessing the impact of a housing project have been referred to in earlier chapters, so this analysis will start by assuming that a standard unit for housing services can be defined (i.e. a room) and that a demand curve for housing measured in terms of this standard unit can be identified. The problems involved in relaxing this assumption will be discussed after examining the nature of housing benefits in a variety of different circumstances. The discussion is organised in terms of a series of adjustments to an initial estimate of project benefits, though the implementation of the method may not involve all of these adjustments. One important point to note is that the question of who owns the housing produced by the project plays an important role in the calculations. The analysis will start by assuming that the housing is owned by a government body which lets the project housing to tenants and receives the full value of the rents paid by the tenants.

Initial estimate

Consider Figure 13.1 which shows the demand curve for rooms in terms of their price (monthly rent). Before the project the total stock of rooms is H_0 and the market rent is r_0 per room per month. As a result of the

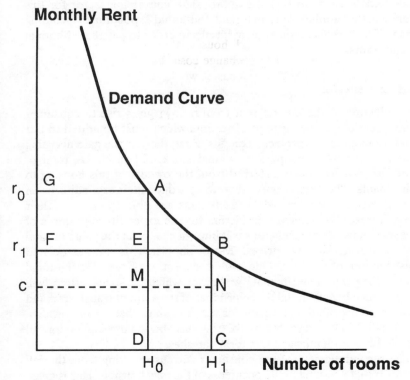

Figure 13.1 The benefits of housing projects

project the housing stock increases to H_1 rooms and the rent falls to r_1. The pre- and post-project combinations of housing stock and rent level are shown as points A and B on the overall demand curve for housing. The total rental revenue for the additional housing is given by the area BCDE, i.e. r_1x where $x = H_1 - H_0$, so that this provides our initial estimate of the willingness to pay for the housing services produced by the project.

Conversion to shadow prices

The initial estimate is expressed in terms of market prices so it (and all of the later adjustments) must be converted to shadow prices before comparing the project's benefits with its costs. The key question is how the revenue generated by the project's housing affects the rest of the economy. Assuming that household budgets are fixed, the project revenue is offset by a reduction in household expenditure on other goods and services, so that the appropriate conversion factor for this revenue is that applicable to marginal household expenditure; including savings if households reduce their savings in order to pay the additional housing rents.

Consumer surplus

The reduction in the market rent from r_0 to r_1 gives rise to consumer surplus accruing to two groups of tenants which must be added to our initial estimate of the project's benefits. First, there is the gain accruing to households who occupy the original stock of housing. A total of AEFG or $(r_0 - r_1)H_0$ is transferred from the owners of this housing to their tenants. The owners must, therefore, reduce their expenditures on other goods and services while the tenants are able to increase their expenditures. If the conversion factors for the expenditure patterns of the two groups are different, this will imply a change in the value of total expenditure at shadow prices. More important is the associated redistribution of income. One would expect that the distributional weights for owners and tenants will differ which, at social prices, will imply a social benefit equal to the product of the amount transferred and the average distributional weight for tenants minus that for the owners. When efficiency prices are used, the distributional weights for all households are identical so this term vanishes.

The second element of the consumer surplus resulting from the fall in rents is that accruing to the occupiers of the new housing. This is given by the area ABE which is the standard consumer surplus 'triangle'. It is equivalent to an increase in the real income of the project's tenants without any offsetting expenditure, so that it should be multiplied by their distributional weight to convert it to shadow prices.

Private landlords, owner-occupiers and below-market rents

Suppose that the project housing is built or purchased by landlords who let it to tenants. The rent revenue generated by the project accrues to these landlords while the government receives some lower sum representing either the site rents for project plots or the payments made by the purchasers of the housing. The precise adjustments which must be made to the initial estimate of benefits will depend upon the circumstances of the project, so their general character will be illustrated by dealing with one example. In Figure 13.1 we assume that the government sells the new housing at a charge of c per room per month to purchasers who are able to earn the rent r_1. Of the total rent revenue $BCDE$ the amount $NCDM$ accrues to the government as before and requires no further adjustment. The amount $BNME$ or $(r_1 - c)x$ represents the net income accruing to the landlords who have purchased the project housing rather than to the government. This income will enable landlords to increase their expenditure but the cost of the resources absorbed by the extra expenditure is partly offset by the benefit associated with increasing the real income of this group. Thus, we must adjust our initial estimate of the project benefits by subtracting the product of the landlords' net income and of the difference between the conversion factor for their expenditure and the distributional weight for such households.

The same adjustment must be made if the purchasers of the project housing occupy it themselves rather than letting it to tenants. In this case, the resulting estimate of the project benefits for the new housing can be rearranged by noting that the revenue accruing to the government amounts to $NCDM$ or cx, while the consumer surplus or income benefit for the occupiers of the housing is $ABNM$. These components are converted to shadow prices by multiplying the first by the conversion factor for marginal household expenditure and the second by the average distributional weight for the households occupying the project housing. This rearrangement is helpful because it is identical to the estimate of benefits which would be obtained if the government were to charge below-market rents for the project's housing. In that case, c would represent the rent actually charged for the project housing while r_1 is the market rent paid by tenants in the original housing stock after the project is completed.

Housebuilding adjustments

So far it has been assumed that the increase in the housing stock from H_0 to H_1 is equal to x, the amount of new housing built for the project. However, it is quite possible that the project may affect the general level of housebuilding activity in the towns or cities in which it takes place. In that case the calculations should be based on the forecast net increase in the housing stock rather than on the amount of housing built for the project. One might expect that this would be less than x on the grounds that the reduction in rents would reduce the level of non-project housebuilding. However, for Tanzania, analysis of the impact of government schemes to upgrade settlements and to offer site and service plots in Dar es Salaam suggested that other factors might lead to an increase in housebuilding. The crucial point was that the government schemes provided official recognition to informal settlements which meant that the owners of informal housing in the areas concerned had much greater security of tenure over the land that they were occupying. As a result these owners were willing to invest in improving the size and quality of their houses, so that the increase in housing supply due to the reduction in uncertainty outweighed any reduction due to the fall in rents.

Large projects

The approximations which underpin the estimation procedures described above are appropriate for small projects, that is for projects which are not expected to cause a significant reduction in the market rent level. They will tend to overstate the decline in rents and thus the redistribution of income from landlords to tenants for larger projects, since the slope of the demand curve will tend to become flatter as the size of the project increases. A simple method of correction would be to adjust the elasticity of the demand curve used in calculating the extent of the decline in rents, but it is difficult to be systematic in determining the magnitude of the appropriate adjustment.

Even if this problem can be surmounted, large projects present another difficulty. Cost-benefit analyses usually assume that the changes in real income experienced by those affected by the project are relatively small. This implies that the marginal utility of income is constant for each household which may not be valid in large projects. It is possible to avoid such an assumption by basing the analysis on a full model of the link between the housing model and household welfare provided that sufficiently good data are available. The benefits of the project may then be calculated by estimating the average changes in expenditure levels and either the compensating variation or the equivalent variation for the various groups of households affected by the project. It is not possible to give the details of this approach here so the interested reader should refer to a text on welfare economics such as Boadway and Bruce (1984). Tax reforms and public pricing policies raise similar issues (see Stern, 1987; and Katz, 1987).

External benefits

The discussion has focused on the benefits which are directly associated with the project's impact on the housing market and with the community's willingness to pay for additional housing. It is possible that improvements in the quality and supply of housing may generate a range of external benefits which are not reflected in rents and the demand for housing. For example, squatter upgrading schemes which provide proper sanitation and less crowded accommodation may make a significant contribution to the health of the residents, thus increasing their welfare, productivity and income levels. The project analyst faces considerable difficulties in treating such effects because it is extremely difficult to define the dividing line between the direct and the external benefits of a project as consideration of this example will illustrate.

To the extent that households are aware of the health benefits of better sanitation and are willing to pay for them via higher rents or charges for piped water and sewerage, these benefits are not external because they are reflected in the information used to compute the direct benefits of the project. In both Kenya and Tanzania there was little doubt that the inhabitants were willing to pay significant fees for the provision of individual piped water connections. Their valuation of individual sewerage connections seemed to be rather less but, if we take water and sewer connections together on the grounds that the provision of piped water without sewers in urban areas leads to severe environmental problems, households' willingness to pay for both greatly exceeded the costs of supplying these services. Thus, the benefits of improved sanitation are recognised and internalised by households, so that no reference to external benefits is required in order to justify the investment required to supply water and sewerage services.

The same problem arises when considering the external benefits of better housing. Households generally recognise the advantages to themselves of reducing overcrowding or improving the quality of their accommodation and are willing to pay for such improvements. However, if the benefits of less crowded and better accommodation as measured in this way are not large enough to pay the additional costs, one should not conclude that this is a symptom of some kind of market failure and that the external benefits of housing improvement should justify the project. Whatever the views of project analysts and other outsiders in such a case, households may reasonably make the judgement that, given their income levels and expenditure priorities, the benefits of better housing do not justify the additional costs involved. It would only be appropriate to allow for additional external benefits when appraising a housing project in such circumstances if there is clear empirical evidence that better housing will confer identifiable benefits on households who are not affected by the project except, perhaps, via any change in rents. The author's interpretation of the literature on the external benefits of housing improvements is that no one has so far made a convincing case that these benefits are likely to be important relative to the direct benefits as measured by households' willingness to pay for better housing.

The suggestion that the external benefits of low income housing projects may be substantial tends to arise from a confusion about the reasons for the low value placed on better housing by households. Such valuations are not the result of a failure to appreciate the benefits of housing improvement but rather they reflect the fact that the households concerned are poor and have other important priorities. The outsider's judgement that households ought to attach a higher value to housing improvement may, therefore, be the expression of a judgement about the importance of relieving poverty or of the view that housing is a 'merit' good. The government's attitude to poverty and the distribution of income is embodied in the distributional weights used in our calculations, so that assigning a greater social priority to relieving poverty would increase the social benefits of a low income housing project by increasing distributional weights for the real income gains accruing to the households affected. However, housing projects may not be particularly effective methods of improving the distribution of income by comparison with alternative policies which increase employment opportunities and real incomes more directly.

The argument that housing is a 'merit' good involves the paternalistic judgement that a household's valuation of its housing should be replaced by a higher valuation determined by an outsider and leads naturally to policies which provide benefits tied specifically to the consumption of housing – i.e. housing subsidies conditional on a minimum consumption level. There is no clear basis for such a substitution of valuations so that economists are not inclined to accept that it is appropriate unless the arguments for treating something as a 'merit' good are extremely strong. This does not seem to be the case for housing, so that analysis of housing projects should rely upon the valuations of households as expressed via their willingness to pay for housing services of various kinds.

Finally, we should return to the question of housing quality and the diversity of housing services that may be encompassed by a single housing project which was finessed earlier by assuming that we can define the demand and supply of housing in terms of a standard unit, the room. Relaxing this assumption can raise potentially intractable problems which are much too large to be dealt with here. In general, cost-benefit analyses of housing tend to rely upon the hedonic pricing technique (see Deaton and Muellbauer, 1980: ch. 10.2; Hufschmidt *et al.*, 1983; Follain and Jiminez 1985; and Chapter 10 in this volume) which assumes that there is a unified market for housing characteristics so that households' willingness to pay for a particular type of housing can be expressed as the sum of the hedonic prices for the separate characteristics which define the specific house type. Provided that a uniform set of hedonic prices for housing characteristics can be identified for the housing market under investigation these may be used to aggregate different kinds of houses and housing services and to express them in terms of a standard unit as assumed above. If the hedonic pricing technique cannot be applied, then it is necessary to analyse the impact of the project on the separate sub-markets for different kinds of housing on the assumption that they are not perfectly integrated. This requires large amounts of data and presents considerable problems in specifying an appropriate economic model, so that it should be avoided if at all possible.

The Kenyan and Tanzanian studies both relied upon the hedonic pricing model in order to generate separate estimates of a household's willingness to pay for basic and higher quality accommodation, services such as communal or individual water connections and sanitation, and general neighbourhood characteristics including better roads and streetlighting. By using such hedonic prices it was possible to evaluate the separate components of the project in order to assess the costs and benefits of devoting more or less resources to the provision of different types of housing or of plots with different levels of services or to the improvement of existing informal and squatter settlements.

COST-BENEFIT ANALYSIS IN PRACTICE

As an illustration of the ways in which cost-benefit analysis may be used to evaluate specific projects and to compare alternative policies or uses of resources, the results of two studies of housing projects in Kenya and

Tanzania are considered. Details of the manner in which the methods described above were applied to one of the Kenyan proposals are given in Hughes (1976), so the focus here is on the interpretation of the results of the evaluations.

Nairobi, Kenya

In 1973 the Government of Kenya and Nairobi City Council requested a loan from the World Bank to finance a site and service scheme for 6000 plots in the Dandora area on the outskirts of Nairobi. The intention of the scheme was that it would provide sufficient housing to replace squatter settlements which had developed in the late 1960s in the Mathare Valley area between the city centre and Dandora.

The design of the project was that each plot would be provided with a core services unit consisting of a WC and a shower. Those allocated plots would be required to construct, either him/herself or by employing other workers, a minimum of a permanent two-room house meeting certain prescribed building standards. These standards implied that the houses were to be built with concrete blocks and would thus be rather more expensive than the mud and wattle or timber houses characteristic of informal settlements such as Mathare Valley. A materials loan and technical assistance with designs and building methods were to be offered to plotholders. It was expected that after four years the typical house built on the plots would contain four rooms; the usual occupancy rate in Mathare Valley was one room per household, so that the whole project might provide housing for 24,000 households which would represent an increase of about 10% in Nairobi's total housing stock.

The crucial issue in the design and evaluation of the project concerned the role of landlords and the level of charges for the plots. The official view was that the plots would be allocated to households whose incomes were close to the minimum industrial wage. They were to be encouraged to build houses for their own occupation, perhaps sub-letting one or two rooms, by keeping the plot charges to only that necessary to cover the cost of the project. However, experience both in Mathare Valley and in previous site and service schemes in Nairobi suggested that restrictions on the transfer of plots would not work, so that the plots would be developed by non-resident landlords. In that case it would be sensible to set much higher plot charges as a means of transferring the surplus generated by letting the houses from landlords to the government.

To capture the range of variables which might affect the social value of the project, Table 13.1 presents the results of the project evaluation for various combinations of assumptions. The distinction between the

Table 13.1 Evaluation of the Dandora Site and Services Project, Nairobi: present values in K£ million at shadow prices

	High level services		Low level services	
	Resident	Landlord	Resident	Landlord
Total project cost	5.52	6.14	4.94	5.56
Net present value – excluding redistribution from landlords to tenants				
Planned plot charge	1.51	0.89	1.80	1.06
Zero plot charge	–0.01	–1.18	0.61	–0.56
Maximum plot charge	2.24	1.88	2.86	2.50
Net present value – including redistribution from landlords to tenants				
Planned plot charge	5.00	4.05	5.29	4.22
Zero plot charge	3.48	1.98	4.09	2.61
Maximum plot charge	5.73	5.04	6.34	5.66
Financial evaluation – net present value at market prices				
Planned plot charge	0.40	0.40	0.39	0.39
Zero plot charge	–3.27	–3.27	–2.51	–2.51
Maximum plot charge	2.17	2.17	2.95	2.95

high and low levels of services concerns the extent of the infrastructure provided as part of the project. The 'resident' estimates assume that all of the plots would be occupied by resident households sub-letting rooms, while the 'landlord' estimates assume that all of the plots would be acquired by non-resident landlords. Total costs are higher in the latter case because non-resident landlords incur management costs which are not imputed to residents. For each combination of assumptions the project was then evaluated on the basis of (a) the planned set of plot charges, (b) no plot charges since it was quite possible that actual collection rates for the plot charges might be very low, and (c) the maximum charge which the City Council might be able to levy.

The results of the evaluations are presented in terms of the net present value of the project at 1973 shadow prices. This was computed using the estimated accounting rate of interest of 10 per cent for Kenya to discount future costs and benefits to the 1973 base period. In view of the scale of the project and the consequent uncertainty about its impact on the general level of rents in Nairobi, the net present value of the project was calculated in two ways. The first, method A, focused entirely on the new housing provided by the project by excluding the effect of the transfer from landlords to tenants as a result of the decline in the average rent for the original housing stock, i.e. the area *AEFG* discussed in the section on consumer surplus. The second, method B, includes this element and thus leads to much higher estimates of the net present value of the project, since the figures in the table show that the general transfer from landlords to tenants accounts for approximately one-third of the total benefits of the project.

The main evaluations of the project are those given in the row 'the net present value including redistribution from landlords to tenants' for 'the planned plot charge'. They show that the project was clearly a good one with a high social rate of return. Even if all of the plots were acquired by non-resident landlords who failed to pay any plot charges, it would still have a positive net present value, though it should be noted that this takes no account of the cost of raising government revenue from the rest of the economy to repay the loans required to finance the project. However, the results show also that the net value of the project would be increased by charging the maximum feasible plot charge which would enable the government to capture most of the surplus which might otherwise have accrued to plotholders.

The differences between the net present values of the project for resident plotholders and non-resident landlords are considerably smaller for the maximum plot charge than for the planned or zero plot charges. The plot charge, up to this level, serves as a tax transferring income from the plotholder or landlord to the government, so that the effect of the transfer on total project benefits depends upon the distributional weight given to plotholders or landlords relative to the numeraire value of 1.0 for government revenue for the Little–Mirrlees system of shadow prices. It may seem inequitable to use plot charges for a low income housing project as a form of taxation, but it must be remembered that they represent a classic non-distortionary tax on a pure rent; while the revenue raised in this way could be used to fund further housing projects or other projects designed to benefit poor households. The main distributional benefit of the project arises because of its impact on the general level of rents in the city rather than from the income accruing to plotholders, so that the government should set out to maximise the volume of new housebuilding rather than the value of transfers made to individual plotholders.

The analysis of the Dandora project was extended by considering alternative methods of developing housing on the same land. To avoid the problems involved in evaluating large projects, ones which are designed to provide 1000 plots or houses each will be examined. These cover the range from site and service schemes with minimal facilities and mud houses to the building of three and four room owner-occupied houses combined with the provision of the requisite mortgage finance. The latter scheme was based on a project being carried out by the Commonwealth Development Corporation in neighbouring part of

Table 13.2 Evaluation of alternative housing projects for Nairobi: present values in K£ million at shadow prices

	Total cost	Net present value A (excl. redistribution)		Net present value B (incl. redistribution)	
Plot charge/rent:		Official	Maximum	Official	Maximum
A. Site and service schemes for 1000 plots with 4 rooms per plot					
1. Swahili mud houses with pit latrines and free communal water					
	0.76	0.38	0.62	0.85	1.09
2. Timber rooms with communal toilets and water					
	0.99	0.28	0.61	0.86	1.19
3. Concrete block rooms with toilets and water on each plot					
	1.01	0.40	0.76	1.05	1.41
B. Council housing schemes for 1000 plots					
1. Concrete block units of 4 rooms with toilets and water on each plot					
Low income tenants	0.76	0.66	0.91	1.35	1.61
Middle income tenants	0.76	0.48	0.83	1.11	1.46
2. 3-room detached houses					
Low income tenants	1.03	0.13	0.39	0.72	0.99
Middle income tenants	1.03	−0.09	0.29	0.47	0.84
C. Mortgage housing scheme: 500 3 room houses + 500 4 room houses					
	1.59	0.36	0.68	1.15	1.47

Nairobi at the same time as the Dandora project was being designed.

The results of evaluating these alternative projects are shown in Table 13.2. Again, the two estimates of the net present values of the projects (excluding and including the redistributive effect of the decline in rents) are provided, as also are estimates for two different levels of plot charges or rent levels. The council housing schemes assume that the Nairobi City Council builds and lets the housing on the plots to tenants who may choose to sub-let part of the accommodation. The council housing project involving three-room detached houses is essentially identical to the standard City Council housing development for its own employees and other council tenants. It is, therefore, striking that this project is clearly the worst of all of those considered, with a negative net present value for method A when the houses are let to middle income tenants. The low rents charged by the City Council combined with relatively high building costs meant that these schemes tended to benefit the households lucky enough to be allocated one of the houses rather than the community at large.

All of the schemes generate a significant positive net present value when we include the redistributive benefits of the decline in rents due to the provision of additional housing. The main conclusions which emerge from the analysis of the various projects are:

1 The best schemes are those which result in the building of concrete block rooms with services on each plot, whether they are built by plotholders in a site and services scheme or by the City Council for letting to poor households.

2 The two principal sources of social benefits for these housing projects are: (a) the redistributive effect operating through the general level of rents, and (b) the taxation effect linked to the plot charges or house rents and the fees for public services.

3 To maximise the redistributive effect of a project the government should not attempt to isolate it from the operation of the general housing market in the city. Not only are rules concerning sub-letting and rent ceilings likely to be widely flouted, but, in as far as they are effective, they tend to reduce the overall benefits of the scheme. Instead of resulting in a general redistribution from landlords to tenants, such restrictions will confer a privileged position on those allocated plots or houses in the project. The greater the benefits of receiving such an allocation, the more probable it is that the units will be acquired by middle income households or by non-resident landlords.

4 If the project housing is regarded as an addition to the overall market supply, plot charges or house rents should be set close to the maximum level compatible with finding tenants for all of the units. This is the taxation aspect of the project benefits since the charges are equivalent to a land tax falling either on landlords or on households whose income is weighted less highly than government revenue. In the short run the charges for public services can be treated similarly, but this does have the disadvantage of discouraging their use and of encouraging the construction of new private housing with lower or no provision of these services. In as far as the primary external benefits are associated with improved sanitation and better quality water, while the revenue from public services is likely to be modest compared to that from plot charges and rents, the best policy would be to set charges for public services equal to their long run marginal cost.

Dar es Salaam, Tanzania

In 1976 the Tanzanian government requested a loan from the World Bank to finance the upgrading of squatter settlements in Dar es Salaam and four smaller towns together with the provision of surveyed plots with a very low level of infrastructure and services. This discussion focuses on Dar es Salaam, for which the best data were available and which raised issues which complemented the Nairobi study examined above. Squatter upgrading involves the provision of basic infrastructure, such as roads and street lighting, plus water and sewerage services. It is usually necessary to demolish some housing in order to keep the costs of construction to a reasonable level, but neighbourhood and political pressures in Dar es Salaam ensure that the loss of housing has to be minimised though the benefits of upgrading are enthusiastically welcomed.

The nature of squatter upgrading projects with their focus on infrastructure and public services and on the establishment of appropriate land-holding and tenure arrangements meant that they presented quite different problems for the project analyst from those associated with housing projects such as those in Nairobi. There were also a number of specific features of the Dar es Salaam housing market in 1976 which influenced the design of the project and the results of the project evaluation:

1 The general level of rents was low and had fallen by nearly 40 per cent in real terms over the previous three years because of a slow down in migration and the growth of housing demand in the capital owing to Tanzania's economic difficulties. Relative to the cost of house building rents had fallen so far that it was not possible to cover even the interest and depreciation on new houses on the basis of a reasonable private rate of return. As a result, the level of house building had fallen very sharply, so that the average growth of squatter areas had been 20 per cent p.a. during 1969–72 but only 5 per cent p.a. in 1972–75. At the prevailing level of rents, it was impossible to justify any project which relied upon the provision of new housing for its primary benefits. However, in the medium and longer term, rents had to rise in order to yield an adequate return on new house building. On this basis it was possible to estimate a long term level of rents in unimproved squatter areas, while the differentials between rents in surveyed plot areas, improved squatter areas, and unimproved squatter areas, were calculated from information on cost, locational, and infrastructure-related differentials in rents between different areas of the city.

2   By examining aerial photographs of different parts of the city, it was possible to demonstrate that the rate of new house building and of housing improvement was significantly higher in improved squatter areas than in unimproved areas. It was estimated that the growth rates of the housing stock would be 8 per cent p.a. and 5 per cent p.a. in improved and unimproved areas respectively, subject to limits on the overall density of housing in both types of area. After the initial period of building, an even faster growth rate of 10 per cent p.a. for the housing stock in surveyed plot areas was anticipated from the addition of extra accommodation and infilling with new units.

3   Surveys of housing quality in different areas of the city revealed that the average quality of new house building was significantly higher than the average quality of the existing stock with much higher proportions of permanent concrete block houses. Further, improved squatter areas with better infrastructure and services experienced a more rapid rate of housing improvement than did unimproved areas.

4   Analysis of rents showed that households were willing to pay a significant rent premium for a room in a house with water and sanitation on the plot, whereas the premium for access to communal water supply and sanitation within 100 metres of a house was very small. As a result, the government's intention to provide water supply and sewerage with communal water taps, washing and toilet facilities, could not be justified in terms of the community's willingness to pay for these services. The provision of private water and sewerage connections was probably justified on the basis of their direct benefits, though it is very difficult to estimate the costs of laying pipes and connections in existing squatter areas because this depends on the amount of compensation paid to residents whose houses have to be demolished in the process.

The implications of these features of the Dar es Salaam housing market were not encouraging for the evaluation of the proposed project. The key question should have been one of timing and design. The low level of rents in 1976 meant that any project started at that time would have a negative net present value if it relied upon the provision of additional housing for the major part of its benefits. However, rents in 1976 were too low to sustain house building and housing improvement at the level required to meet the projected rate of population growth for the city, so it was reasonable to evaluate the project for future implementation when rents had increased to what were estimated as their equilibrium level.

Table 13.3 illustrates the relative influence of the various factors on

Table 13.3 Evaluation of squatter upgrading and surveyed plots in Dar-es-Salaam

		Internal rates of return (percentages at 1976 shadow prices)
1.	Squatter upgrading and surveyed plots	
	Main evaluation for standard parameters	17.7
	Evaluations for alternative assumptions :	
	(a) Identical rates of house building in improved & unimproved areas	17.5
	(b) Identical rates & composition of house-building in improved & unimproved areas	17.5
	(c) Lower rent differentials between house types	14.9
	(d) Lower rent levels and differentials between house types	11.8
	(e) Lower income weight for residents of squatter areas	12.7
2.	Squatter upgrading alone	
	Main evaluation for standard parameters	14.3

the results of the evaluation of the Dar es Salaam housing component of the project including the upgrading of squatter areas and the provision of surveyed plots. The evaluations are presented in terms of the internal rates of return for the alternative project assumptions because total cost and benefit estimates are affected by some of the assumptions in ways which make comparisons difficult. There is a direct relationship between net present values, the discount rate and internal rate of return so that, in the Kenyan examples above, any project with a positive net present value would have an internal rate of return in excess of the discount rate which was 10 per cent p.a. It was felt that a discount rate of between 8 per cent and 10 per cent p.a. was appropriate, so that projects earning an internal rate of return of less than 10 per cent were either marginal or unprofitable in social terms.

The comparisons show that it is the willingness of households to pay for housing in terms of the general level of rents and the quality differentials between rents for the various house types which were most important in determining the value of the project. The second important factor was the distributional weight assigned to real income benefits accruing to residents of the squatter areas. The government was acting as if it gave as much weight to increasing the real income of the average squatter household as to its own revenue, because it was making little effort to collect taxes, land rents, and service charges from such households and it would pay full compensation to displaced squatters if they lost buildings or crops as a result of upgrading schemes. It is doubtful whether the government was justified in assigning such a high weight to the real income of squatter households in Dar es Salaam because they were substantially better off than the average for the smaller towns and, even more, for rural areas.

The combination of lower rents and a lower income weight for squatter households clearly meant that the project was unlikely to generate a satisfactory rate of return in 1976, though the rate of return would have increased over time as rents increased towards their long term equilibrium level.

Table 13.3 also compares the return on the combination of squatter upgrading and surveyed plots with that on squatter upgrading alone. This is less straightforward than it might seem because much of the investment in infrastructure was shared between the two components. However, it was clear that the surveyed plots could not proceed without the squatter upgrading, so the latter was evaluated on the assumption that all costs should be assigned to the squatter upgrading other than those directly linked to the surveyed plots. The analysis showed a substantially lower rate of return for the squatter upgrading component on it own, so that the sharing of costs from combining the two elements leads to a significant increase in the return on the project.

By the time the negative evaluation of the project emerged the project preparation phase was almost complete. Both the government and the World Bank were, understandably, unwilling to postpone the project on the basis of the conclusions of a single evaluation. Instead, they chose to rely on the fact that the project could be justified on the longer term perspective. Uncertainties about the recovery of the Tanzanian economy and the prospective rate of growth of Dar es Salaam meant that it was not possible to estimate how long the project should be delayed before it would become socially profitable. It was, in any case, likely that implementation of the project might be delayed because of staff and other resource constraints, so that any attempt to choose an optimal timing for the project was unlikely to work in practice.

CONCLUSION

These two case studies have shown that cost-benefit analysis should not just be viewed as a tool designed to provide a yes–no answer as to whether it is worth going ahead with a project. Instead, it provides a consistent framework for examining many aspects of project design and policy formulation. In the Kenyan study it was possible to use the method to compare alternative types of house building and levels of service provision as well as institutional and policy issues such as the setting of council rents and plot charges and the impact of regulations concerning the allocation of plots and the role of landlords in the development of site and service schemes. The Tanzanian study involved consideration of the appropriate priorities in the design of programmes for upgrading squatter areas and the relationship between government provision of basic infrastructure and private development of the housing stock. This case also showed that the timing of a project may be crucial to whether it can be justified or not.

Finally, the most important lesson that should be learned from both studies is that cost-benefit analysis should not be left to the last stages of project preparation when the design and timing of the project cannot easily be changed. Results of the kind outlined above should play an important role in deciding upon the design of project components and the balance between these components. This would mean that unpleasant surprises at the results of a project evaluation, such as occurred in Tanzania, should not arise and it would facilitate the formulation of projects likely to produce the best use of the resources involved.

ACKNOWLEDGEMENTS

I am grateful to the editors for their indulgence in coping with the delays which have accompanied the writing of this chapter. Parts of the chapter are based on studies originally funded by HM Government Overseas Development Administration (in Kenya) and the World Bank (in Tanzania). I am grateful to both organisations and also to the many individuals and organisations who provided the basic data on which the work was based for their support and assistance. Finally, I am solely responsible for the analysis and views expressed in this chapter.

14 Methods of analysis and policy

Kenneth G. Willis and A. Graham Tipple

The rigorous application of techniques and methods of analysis, such as those outlined in this book, to housing research and policy decisions will improve judgements on housing issues. The purpose of applying methodology is to improve consumer satisfaction with housing, and promote economic efficiency and greater equity (fairness) in housing policy. It is hoped that substantial improvements can be made to the design and pricing of housing projects and to the operation of housing markets by making the analysis of housing projects and policies more technical, objective, and informed, and less subjective, political and ideological.

Few commentators on the state of the housing market in any country express satisfaction with the outcome of their respective country's housing policies. Irrespective of their ideological positions, subject or discipline orientation, researchers tend to agree that past policy has failed. After decades of state intervention in housing in most countries, the physical manifestations of housing problems still exist: shortage of supply leading to homelessness, long waiting times to acquire houses, overcrowded households and squatting; houses in poor condition, lacking amenities and highly priced in relation to income.

Public housing projects and housing policy may be seen as a response to market failure. Market failures (monopoly, technological externalities, uncertainty and inequity) are traditional reasons in welfare economics for government intervention (Willis, 1980). Externalities in housing might encompass public health concerns and urban blight because of prisoner's dilemma problems (giving rise to neighbourhood renewal schemes). Equity is commonly thought of in terms of income distribution (housing benefits, public housing for low income groups). However, the results of housing policy do not closely accord with such welfare maximising objectives. Many aspects of housing policy tend to be inefficient and inequitable. For example, owner occupiers receive large tax benefits in terms of interest relief on housing loans, most of which is captured by upper income groups (Robinson, 1981; Sawhill, 1990). Private rented accommodation is dilapidated and in short supply due to the excessive taxation and rent controls (Willis, Malpezzi and Tipple, 1990). High quality public housing is supplied to people who cannot afford to pay rents which cover the costs of such housing, and hence such housing must be subsidised, though the utility households derive from such housing is less than the subsidy and hence is inefficient (Watson and Holman, 1977: 107–9; Malpezzi and Mayo, 1985).

Ninety-nine per cent of all economists, for example, would point to tax relief on mortgage interest payments as being inefficient (mainly affecting the price rather than the supply and ability to pay for housing) and inequitable (most tax benefits go to richer households) and advocate its abolition. Yet such housing policies are perpetuated, because any political party advocating the abolition of this policy would lose votes at the ballot box or in the political market place. Similarly with rent control and many other housing policies, to change is to risk losing votes and elections. Moreover, policies such as rent control, which initially appear to benefit many at a cost to a few (landlords), appeal to the electorate if included in a political manifesto. Even housing policies which only benefit minorities, at the cost of public subsidy, can be electorally profitable if the cost of each policy to all the public is so small (e.g. 1 or 5 cents per year) as to be hardly worth considering when voting. If sufficient such policies can be offered to different minorities without being seen to perceptively increase an individual's tax contribution, votes can be bought and accumulated.

Analysis of expenditure by local housing authorities in Britain has suggested that financial restraints and the marginal cost of funds do not impose severe restrictions on capital spending plans. That is, the social marginal cost of funds schedule is flat for much of its length, before rising steeply where high investment levels cause risk (such as political risk, or risk of failure to obtain revenues). Variations in the social marginal efficiency of investment (need indicators for housing, such as overcrowding, lack of amenities, etc.) constituted the most important variations in capital expenditure on housing (Nicholson and Topham, 1971). However, the attitude and political composition of housing authorities was also noted to be instrumental in investment decisions.

This view of local authority housing expenditure decisions and housing policy in general being determined by a public choice perspective (i. e. politically determined by the voting mechanism rather than solely determined by efficiency and distributional considerations) was subsequently investigated by Ricketts (1981, 1982, 1983). He concluded that results support the theory of regulation: that regulatory processes in many sections of the housing market are not concerned so much with correcting market failure but rather should be seen as being part of a political struggle over the distribution of wealth. Public choice theory reveals the power of interest groups to affect housing decisions. Party and ideological factors affect public subsidies to housing and rent levels set in public housing. Evidence also exists to suggest that electoral security enables politicians to pursue their own policy preferences more easily.

Similar claims have been made about zoning laws in the USA. Zoning restrictions are instituted in an effort to eliminate external diseconomics which the construction of 'undesirable' property features might impose upon other properties in any given zoning district. While zoning can affect property values, evidence suggests that changes in property prices due to zoning are inconsistent with the hypothesis that zoning eliminates the threat of externalities (Avrin, 1977). Conversely, it has been argued that wealth redistribution rather than efficiency is a more useful vehicle for analysing zoning (Goetz and Wofford, 1979). Zoning can be examined from a regulatory framework, and zoning is an example of rent seeking, with all that that implies for efficiency (Krueger, 1974) and economic growth (Agarwala, 1983).

While chapters in this book have concentrated on such issues as efficiency losses or measuring distributional outcomes, whether in financial, economic, social and cultural terms or over time, public choice theory explains why housing policy outcomes so often appear to be arbitrary, inequitable and inefficient. Housing policy will in the future also be arbitrary, inequitable and inefficient if it continues to be determined in the political market place through majority and plurality voting mechanisms. Conversely, if housing policy was less political or politicised, and more technical and in tune with people's cardinal preferences, rather than those expressed through some nominal or ordinal voting system, then society's welfare would be increased and housing markets would become more efficient and equitable.

The techniques and subject matter within this book have concentrated on a positive rather than a normative perspective. Positive statements may be simple or complex, but they are essentially about what *is* the case. Disagreements over positive statements can be resolved by an appeal to the facts: hence the application of each technique to a case study. Normative statements concern what ought to be and stem from ideological, cultural, and religious beliefs, judgements, and preferences. The validity or otherwise of normative statements cannot be settled by an appeal to empirical observation.

The individual chapters in this book have sought to answer such questions as: How far does state-aided housing reflect the needs of its occupants? What are the core elements in a culture which need supporting by housing design and policy and what are merely peripheral? How far do similar policies lead to different outcomes in different cities? Is subsidised lending to finance housing captured by richer households? How much of their income are households willing to pay to own a house? If financial institutions lent to low income groups, would this increase their risks of default by borrowers? Are owner households intrinsically different from renter households? Does overcrowding result from demographic or economic features of the household? How far do rent controls benefit low income groups and do they affect the supply of housing? Is a particular package of housing interventions likely to benefit investors, occupants and the government? If price distortions in the economy were removed, what effect would this have on housing construction? And what is to be gained from keeping plot prices as low as possible in site and service schemes compared with setting higher plot charges?

Concentration on positivism in practice does not preclude the examination of normative statements about what policy *should* or *ought* to be adopted. The assertion that 'rent control ought to be introduced because commands and controls are good policy instruments' is a normative statement. However, it is perfectly acceptable to ask why? What consequences would ensue from rent control? Statements about the consequences of rent control are positive, testable statements. Thus the pursuit of what appears to be a normative position will often turn up positive hypotheses upon which the 'ought' conclusion depends or by which it is judged. The techniques and methods advanced in this book are designed to explore, analyse, and assess such positive statements and hypotheses.

This book has demonstrated that methodologies exist to enable us to go beyond anecdotal evidence, and make housing analysis a more technical and objective science in developing countries. In this way techniques can make a positive contribution to ensuring that housing markets are more efficient and equitable, and that houses are fit for households.

Bibliography

Abu-Lughod, J. (1976) 'The legitimacy of comparisons in comparative urban studies: a theoretical position and an application to North African cities', in J. Walton and L.H. Masotti (eds) *The City in Comparative Perspective*, New York: Halsted Press, pp. 17–39.

Agarwala, R. (1983) *Price Distortions and Growth in Developing Countries*, World Bank Staff Working Papers No. 575, Washington, DC: The World Bank.

Amis, P. (1987) 'Migration, urban poverty and the housing market: the Nairobi case', in J. Eades (ed.) *Migrants, Workers, and the Social Order*, London: Tavistock, pp. 249–68.

Armer, M. and Marsh, R. (eds) (1982) *Comparative Sociological Research in the 1960s and 1970s*, Leiden: E.J. Brill.

Asabere, P.K. (1981) 'The price of urban land in a chiefdom: empirical evidence on a traditional African city, Kumasi', *Journal of Regional Science*, 21, 529–39.

Avrin, M.E. (1977) 'Some economic effects of residential zoning in San Francisco', in G.K. Ingram (ed.) *Residential Location and Urban Housing Markets*, Cambridge Mass.: Ballinger.

Badura, B. (1986) 'Social networks and the quality of life' in D. Frick (ed.) *The Quality of Urban Life (Social, Psychological and Physical Conditions)*, Berlin: de Gruyter, pp. 55–60.

Bamberger, M., Sanyal, B. and Valverde, N. (1982) 'Evaluation of sites and services projects: the experience of Lusaka, Zambia', World Bank Staff Working Papers No. 548, Washington, DC: The World Bank.

Banda, M.C. (1978) 'The de-densification process in George Complex (Monitoring Report)', Lusaka: Lusaka Housing Project Evaluation Team Working Paper 16.

Barber, G.M. (1988) *Elementary Statistics for Geographers*, New York: The Guilford Press.

Barnouv, V. (1979) *Anthropology: a General Introduction*, Chicago, Ill.: Dorsey Press.

Bartholomew, D.J. (1967) *Stochastic Models for Social Processes*, London: John Wiley.

Bechtel, R.B., Marans, R.W. and Michelson, W. (1987) *Methods in Environmental and Behavioral Research*, New York: Van Nostrand Reinhold.

Becker, G.S. (1981) *A Treatise on the Family*, Cambridge, Mass.: Harvard University Press.

Berger, P.L. (1981) 'Speaking to the third world', *Commentary* 72, 4 (October), 29–36.

Berger, P.L. (1983) 'Democracy for everyone?' *Commentary* 76, 3 (September), 31–6.

Bergstrom, J. C., Stoll, J.R. and Randall, A. (1989) 'Information effects in contingent markets', *American Journal of Agricultural Economics* 71, 3 (August), 686–91.

Bertaud, A., Bertaud, M-A. and Wright, J.O. (1981) 'The Bertaud Model: a model for the analysis of alternatives for low-income shelter in the developing world', Urban Development Department Technical Paper No. 2, Washington, DC: The World Bank.

Bertaud, A., Bertaud M-A. and Wright, J.O. (1988) 'Efficiency in land use and infrastructure design: an application of the Bertaud Model'. Discussion Paper INU 17, Washington, DC: The World Bank.

Bishop, R. C., Heberlein, T.A. and Kealy, M.J. (1983) 'Contingent valuation of environmental assets: comparison with a simulated market', *Natural Resources Journal* 23, 2, 619–34.

Blackwood, F. (1983) 'The Performance of Men and Women in Repayment of Mortgage Loans in Jamaica', Report prepared for the Population Council of New York.

BNH (National Housing Bank of Brazil) (1984) *Relatório Anual – 1984*, Rio de Janeiro: BNH.

BNH (1985a) Relatorio de quebra, Região metropolitana de Curitiba, Rio de Janeiro: BNH.

BNH (1985b) Relatorio de quebra, Região metropolitana de Salvador, Rio de Janeiro: BNH.

Boadway, R.W. and Bruce, N. (1984) *Welfare Economics*, Oxford: Blackwell.

Bohm, P. (1972) 'Estimating the demand for public goods: an experiment', *European Economic Review* 3, 2, 111–30.

Boland, L.A. (1979) 'A critique of Friedman's critics', *Journal of Economic Literature* 17, 503–22.

Boyle, K.J. (1989) 'Commodity specification and the framing of contingent valuation questions', *Land Economics* 65, 1 (February), 57–63.

Brazil Report (1981) No. 9.

Brealey, R. and Myers, S. (1981) *Principles of Corporate Finance*, New York: McGraw-Hill.

Breese, G. (ed) (1969) *The City in Newly Developing Countries*, Englewood Cliffs, NJ: Prentice-Hall.

Brookshire, D.S. and Crocker, T.D. (1981) 'The advantages of contingent valuation methods for benefit-cost analysis', *Public Choice* 36, 2: 235–52.

Brookshire, D.S., Schulze, W.D., Thayer, M.A. and D'Arge, R.C. (1982) 'Valuing public goods: a comparison of survey and hedonic approaches', *American Economic Review* 72, 1, 165–77.

Brueggeman, W. (1985) *Federal Rental Housing Production Incentives: Effect on Rents and Investor Returns*, Washington, DC: US Government Accounting Office.

Brunstein, F. (1988) 'Presentacion – Agua y saneamiento en America Latina', *Medio Ambiente y urbanizacion* 23, 1.

BSAJ (Building Societies Association of Jamaica) (1989) *1988 Fact Book*, Kingston: BSAJ.

Buchanan, J.M. (1983) 'Rent seeking, noncompensated transfers, and laws of succession' *Journal of Law and Economics* 24, 71–85.

Bunker, S.G. (1985) *Underdeveloping the Amazon: Extraction, Unequal Exchange, and the Failure of the Modern State*, Urbana: University of Illinois Press.

Burgess, R. (1982) 'Self-help housing advocacy – a curious form of radicalism: a critique of the work of J.F.C. Turner', in P. Ward (ed.) *Self-Help Housing: A Critique*, London: Mansell, pp. 55–97.

Burgess, R. (1985) 'The limits of state self-help housing programmes' *Development and Change* 16, 271–312.

Burns, L, and Grebler, L. (1976) 'Resource allocation to housing investment: a comparative international study', *Economic Development · and Cultural Change* 25, 1 (October), 95–121.

Byatt, I.C.R., Holmans, A.E. and Laidler, D.E.W. (1973) 'Income and demand for housing: some evidence for Great Britain', in M. Parkin and A.R. Nobay (eds) *Essays in Modern Economics* (Proceeding of the Annual Conference of the Association of University Teachers of Economics, Aberystwyth, 1972), London: Longman, pp. 65–84.

Carlson, D.B. and Heinberg, J.D. (1978) 'How housing allowances work: integrated findings from the experimental housing allowance program', Urban Institute Paper on Housing No. 249–3, Washington, DC: The Urban Institute.

Casley, D.J. and Lury, D.A. (1987) *Data Collection in Developing Countries*, London: Oxford University Press.

Casley, D.J. and Kumar, K. (1988) *The Collection, Analysis, and Use of Monitoring and Evaluation Data*, Baltimore, Johns Hopkins University Press for the World Bank.

Chou, Y-L. (1975) *Statistical Analysis*, New York: Holt, Rinehart and Winston.

Churchill, A.A. (1987) 'Rural water supply and sanitation: a time for change', World Bank Discussion Papers 18, Washington, DC: The World Bank.

Cicchetti, C.J., Smith, V.K., Knetsch, J.H. and Patton, R.A. (1972) 'Recreation benefits estimation and forecasting: implications of the identification problem', *Water Resources Research* 8, 4 (August), 840–50.

Clark, W.A.V. and Heskin, A.D. (1982) 'The impact of rent control on tenure discounts and residential mobility', *Land Economics* 58, 1, 109–17.

Cliff, A.D. and Ord, J.K. (1973) *Spatial Autocorrelation*, London: Pion.

Cliff, A.D. and Ord, J.K. (1981) *Spatial Processes: models and applications*, London: Pion.

Cummings, R.G., Brookshire, D.S. and Schulze, W.D. (eds) (1986) *Valuing Environmental Goods: An Assessment of the Contingent Valuation Method*, Totowa, NJ: Rowman and Allanheld.

Datta, G. and Meerman, J. (1980) 'Household income or household income per capita in welfare comparisons', World Bank Staff Working Papers No. 378, Washington, DC: The World Bank.

Davis, O. and Whinston, A. (1961) 'The economics of urban renewal', *Law and Contemporary Problems* 26, 105–17.

Dawes, R.M. (1980) 'You can't systematize human judgement: dyslexia', *New Directions for Methodology in Social and Behavioural Science*, 4, 67–78.

Deaton, A. and Muellbauer, J. (1980) *Economics and Consumer Behaviour*, Cambridge: Cambridge University Press.

Deyak, T.A. and Smith, V.K. (1978) 'Congestion and participation in outdoor recreation: a household production function approach', *Journal of Environmental Economics and Management* 5, 1, 63–80.

Doling, J.F. (1973) 'A two stage model of tenure choice in the housing market', *Urban Studies* 10, 199–211.

de Dombal, F.T., Leaper, D.J., Horrocks, J.C., Staniland, J.R. and McCann, A.P. (1974) 'Human and computer aided diagnosis of abdominal pain: further report with emphasis on performance of clinicians', *British Medical Journal* 1, 376–80.

de Dombal, F.T. (1984) 'Computer aided diagnosis of acute abdominal pain: the British experience', *Revue d'Epidemiologie et de Santé Publique* 32, 50–6.

Draper, N.R. and Smith, H. (1981) *Applied Regression Analysis*, 2nd edn, New York: John Wiley.

Dreze, J. and Stern, N.H. (1987) 'The theory of cost-benefit analysis', in A.A.J. Auerbach and M. Feldstein (eds) *Handbook of Public Economics*, Vol. 2, Amsterdam: North Holland, pp. 909–90.

Duncan, J.S. 'From container of women to status symbol: the impact of social structure on the meaning of the house', in J. Duncan (ed.) *Housing and Identity: Cross Cultural Perspectives*, New York: Holmes and Meier.

Durand, D. (1941) *Risk Elements in Consumer Instalment Financing*, Studies in Consumer Instalment Financing No. 8, Washington, DC: US National Bureau of Economic Research, quoted in G. Tintner (1965) *Econometrics*, New York: John Wiley, p. 98.

Edwards, M. (1982) 'Cities of tenants: renting among the urban poor in Latin America', in A. Gilbert with J.E. Hardoy and R. Ramirez (eds) *Urbanisation in Contemporary Latin America*, Chichester: John Wiley.

Evans, P. B., Rueschemeyer, D. and Skocpol, T. (1985) *Bringing the State Back In*, Cambridge: Cambridge University Press.

Everitt, B.S. and Dunn, G. (1983) *Advanced Methods of Data Exploration and Modelling*, London: Heinemann.

Fathy, H. (1973) *Architecture for the Poor: An Experiment in Rural Egypt*, Chicago, Ill.: University of Chicago Press.

Feber, R. and Hirsch, W.Z. (1982) *Social Experimentation and Economic Policy*, Cambridge: Cambridge University Press.

Fischhoff, B. and Furby, L. (1988) 'Measuring values: a conceptual framework for interpreting transactions with special reference to contingent valuation of visibility', *Journal of Risk and Uncertainty* 1 (June), 147–84.

Fleming, M.C. and Nellis, J.G. (1987) *Spon's House Price Data Book*, London: E. and F.N. Spon.

Fogerty, M.S. (1977) 'Predicting neighbourhood decline within a large central city: an application of discriminant analysis', *Environment and Planning*, A 9, 579–84.

Follain, J. and Jimenez, E. (1985) 'Estimating the demand for housing characteristics: a survey and critique', *Regional Science and Urban Economics* 15, 77–107.

Franck, K. (1985) 'Change: a central but unheralded theme in environmental design research', in S. Klein, R. Wener, and S. Lehman (eds) *Environmental Change/Social Change: Proceedings of EDRA 16*, New York: Environmental Design and Research Association (EDRA).

Freeman, A.M. (1979) 'The hedonic price approach to measuring demand for neighbourhood characteristics', in D. Segal (ed.) *The Economics of Neighbourhood*, New York: Academic Press.

Friedman, J., Jimenez, E. and Mayo, S.K. (1988) 'The demand for tenure security in developing countries', *Journal of Development Economics* 29, 185–98.

Friedmann, J. and Wulff, R. (1976) *The Urban Transition: Comparative Studies of Newly Industrialising Societies*, London: Edward Arnold.

Geiger, P.P. and Davidovich, F.R. (1986) 'The spatial strategies of the state in the political-economic development of Brazil', in A. Scott and M. Storper (eds) *Work, Production, Territory: The Geographical Anatomy of Industrial Capitalism*, London: Allen and Unwin: 281–300.

Gilbert, A.G. (1983) 'The tenants of self-help housing: choice and constraint in the housing markets of less developed countries', *Development and Change* 14, 449–77.

Gilbert, A.G. (1987a) 'Research policy and review 15: From Little Englanders into Big Englanders: thoughts on the relevance of relevant research', *Environment and Planning A* 19, 143–51.

Gilbert, A.G. (1987b) 'Latin America's Urban Poor: shanty dwellers or renters of rooms?' *Cities*, (February), 43–51.

Gilbert, A.G. (1989) *Rental Housing in Developing Countries*, Nairobi: Report to UN Centre for Human Settlements, also published as United Nations (1989) *Strategies for Low-Income Shelter and Services Development: the Rental-Housing Option*, Nairobi, UNCHS (Habitat).

Gilbert, A.G. and Goodman, D. (1976) 'Regional income disparities and economic development', in A.G. Gilbert (ed.) *Development Planning and Spatial Structure*, New York: Wiley.

Gilbert, A.G. and Healey, P. (1985) *The Political Economy of Land*, Aldershot: Gower.

Gilbert, A.G. and Varley, A. (1990) *Landlord and Tenant: Housing the Poor in Urban Mexico*, London: Routledge.

Gilbert, A.G. and Ward, P.M. (1985) *Housing, the State and the Poor: Policy and Practice in Three Latin American Cities*, Cambridge: Cambridge University Press.

Gittinger, J.P. (1982) *Economic Analysis of Agricultural Projects*, Baltimore: Johns Hopkins University Press.

Glaser, D. (1985) 'Who gets probation and parole: a case study versus actuarial decision making', *Crime and Delinquency* 31, 367–78.

Gleaner (Kingston) (1976a) 'Proposed National Housing Trust operations outlined', 17 February, 5.

Gleaner (Kingston) (1976b) 'House approves National Housing Trust', 18 March, 12 & 16.

Gleaner (Kingston) (1984) 'Government feels 80,000 housing units per annum needed', 9 November.

Goetz, M.L. and Wofford, L.E. (1979) 'The motivation for zoning: efficiency or wealth distribution', *Land Economics* 55: 472–85.

Golding, B. (1982) *A National Housing Policy for Jamaica*, Kingston: MOC(H).

Graham, D. and Pollard, S. (1982) 'Credit project transformed into an ad hoc income transfer programme', *Social and Economic Studies* 32 (March), 63–80.

Graves, P., Murdoch, J.C., Thayer, M.A. and Waldman, D. (1988) 'The robustness of hedonic price estimation: urban air quality', *Land Economics* 64, 220–33.

Grose, R.N. (1979) *Squatting and the Geography of Class Conflict: Limits to Housing Autonomy in Jamaica*, Ithaca NY: Cornell University Program on International Studies in Planning.

Hamdi, N. (1978) '"PSSHAK" – Adelaide Road', *Open House* 3, 2, 18–22.

Hammond, K.R. (1978) 'Towards increasing competence of thought in public policy formulation', in K.R. Hammond (ed.) *Judgement and Decision in Public Policy Formation*, Boulder, Col.: Westview Press.

Hammond, P. (1988) 'Principles for evaluating public sector projects', in P. Hare (ed.) *Surveys in Public Sector Economics*, Oxford: Blackwell, pp. 15–44.

Hamnett, M. P., Porter, D. J., Singh, A. and Kumar, K. (1984) *Ethics, Politics, and International Social Science Research: from Critique to Praxis*, Honolulu: University of Hawaii Press.

Hansen, K.T. (1982) 'Lusaka's squatters: past and present', *African Studies Review*, 15, 2/3, 117–36.

Hardie, G.J. (1980) 'Tswana Design of House and Settlement: Continuity and Change in Expressive Space', Unpublished PhD Dissertation, Boston University.

Hardie, G.J. (1982) 'The dynamics of the internal organization of the traditional tribal capital Mochudi', in R.R. Hitchcock and M.R. Smith (eds) *Settlement in Botswana*, Johannesburg: Heinemann Educational Books, pp. 205–19.

Hardie, G.J. (1985a) 'Continuity and change in Tswana's house and settlement form', in I. Altman and C. Werner, (eds) *Home Environments, Human Behaviour and Environments*, New York: Plenum Press.

Hardie, G.J. (1985b) 'Tswana concepts of placemaking', in K. Dovey, G. Downton and G. Missingham (eds) *Place and Placemaking*, Proceedings of the PAPER 85 Conference, pp. 137–43.

Hardie, G.J. (1986) 'Continuity and change in expressive space', in D. Saile (ed.), *Architecture in Cultural Change (Essays in Built Form and Culture Research)*, Lawrence: University of Kansas.

Hardie, G.J. (1988) 'Community participation based on three-dimensional simulation models', *Design Studies* 9, 1 (January), 56–61.

Hardie, G.J. (1989) 'Environment and behaviour research in developing countries', in E.H. Zube and G.T. Moore (eds) *Advances in Environment, Behaviour and Design*, Vol. 2, New York: Plenum Press.

Hardie, G.J. and Hart, T. (1985) 'Physical planning and community involvement: an experiment using participation techniques in Mangaung, Bloemfontein', *Housing in South Africa* 4–5 (January).

Hardie, G.J. and Hart, T. (1989) 'Politics, culture and the built form: user reaction to the privatization of state housing in South Africa', in S. Low and E. Chambers (eds) *Housing, Culture and Design: A Comparative Perspective*, Philadelphia: University of Pennsylvania.

Hardie, G.J., Hart, T. and Theart, C. (1986) 'Community involvement in planning', *Town and Regional Planning* 21 (April), 9–13.

Hardoy, J. and Satterthwaite, D. (1986) 'Shelter, infrastructure, and services in third world cities', *Habitat International* 10, 3, 245–84.

Healey, P. (1974) 'Planning and change', *Progress in Planning* 2, 143–237.

Hoenderdos, W., Van Lindert, P. and Verkoren, O. (1983) 'Residential mobility, occupational changes and self-help housing in Latin American cities: first impressions from a current research programme', *Tijdschrift voor Economische et Sociale Geografie* 74, 376–86.

Hublin, A. (1989) 'Analysing aerial photographs of traditional Maroon settlements', *Traditional Dwellings and Settlements Review* 1, 1 (Fall), 83–102.

HUDCO (1982) *Computer Based Design and Analysis of Affordable Shelter: HUDCO Model*, New Delhi: HUDCO.

Hufschmidt, M., James, D.E., Meister, A.D., Bower, B.T. and Dixon, J.A. (1983) *Environment, Natural Systems and Development: An Economic Valuation Guide*, Baltimore: Johns Hopkins University Press.

Hughes, G.A. (1976) 'Low income housing: a Kenyan case study', in I.M.D. Little and M.F.G. Scott (eds), *Using Shadow Prices*, London: Heinemann Educational Books, pp. 43–87.

Hughes, G.A. (1986) 'Shadow prices and conversion factors', *Project Appraisal* 1: 106–20.

Jaffee, A. (1985) *Analyzing Real Estate Investment Decision Using Lotus 1–2–3* Clearwater, Fla.: Reston Stuart.

Jagannathan, N.V. and Halder, A. (1987) 'Income-housing linkages: a case of the pavement dwellers of Calcutta', Paper presented at the International Seminar on Income and Housing in Third World Urban Development, December, HUDCO, New Delhi.

de Jesus, C.M. (1962) *Beyond all Pity: The Diary of Carolina Maria de Jesus*, London: Panther.

Johnson, R.J. (1976) *Classification in Geography*, Norwich: Geo Abstracts.

Johnson, R.J. (1978) *Multivariate Statistical Analysis in Geography*. London: Longman.

Jones, E., Webber, M. and Turner, M.A. (1987) *Jamaica Shelter Sector Strategy, Phase 1 – Final Report*, USAID Sponsored 'U.I. Project 3666–02', Washington, DC: The Urban Institute.

Jules-Rosette, B. (1981) *Symbols of Change: Urban Transition in a Zambian Community*, New Jersey: Ablex Publishing Corporation.

Kahneman, D. and Knetsch, J. (1990) 'Valuing public goods: the purchase of moral satisfaction', *Journal of Environmental Economics and Management* (in press).

Katz, M. (1987) 'Pricing publicly supplied goods and services', in D.M.G. Newbery and N.H. Stern (eds) *The Theory of Taxation for Developing Countries*, Oxford: Oxford University Press, pp. 560–88

Kendall, M. (1975) *Multivariate Statistics*. London: Charles Griffin.

Kennedy, P. (1979) *A Guide to Econometrics*, Oxford: Martin Robertson.

Kent, S. (1984) *Analysing Activity Areas (An Ethnoarchaeological Study of the Use of Space)*, Albuquerque: University of New Mexico Press.

Kish, L. (1965) *Survey Sampling*, New York: John Wiley.

Klak, T. (1990) 'Spatially and socially progressive state policy and programs: the case of Brazil's National Housing Bank', *Annals of the Association of American Geographers* (in press).

Kleinmuntz, B. and Szucko, J.J. (1984) 'Lie detection in ancient and modern times: a call for contemporary scientific study', *American Psychologist* 39, 766–76.

Kohn, M.L. (ed.) (1989a) *Cross-national Research in Sociology*, New York: Sage Publications.

Kohn, M.L. (1989b) Cross-national research as an analytic strategy', in M.L. Kohn (ed.) *Cross-national Research in Sociology*, New York: Sage Publications: 77–102.

Kolenda, P.M. (1968) 'Region, caste, and family structure: a comparative study of Indian 'joint' family', in M. Singer and B. Cohn (eds) *Structure and Change in Indian Society*, Chicago: Aldine.

Kornblum, W. and Beshers J. (1989) 'White ethnicity: ecological dimensions' in J.H. Mollenkopf (ed.) *Power, Culture and Place: Essays on New York City*, New York: Russell Sage Foundation, pp. 201–21.

Koutsoyiannis, A. (1977) *Theory of Econometrics*, 2nd edn, London: Macmillan.

Krueger, A.O. (1974) 'The political economy of the rent seeking society', *American Economic Review* 64, 291–303.

Kumar, O. (1987) *Sites and Services in Urban Housing in India*, New Delhi: Ess Ess Publications.

Lal, D. (1978) *Men or Machines*, Geneva: International Labour Office.

Lal, D. (1980) *Prices for Planning*, London: Heinemann Educational.

Landeau, J-F. (1985) *Tunisia: the Formal Housing Finance System*, Washington, DC: The World Bank.

Landeau, J-F. (1987) 'Tunisia: a case study in analysing the affordability of mortgage loans', *African Urban Quarterly* 2, 3 (August), 223–33.

Leeds, A. and Leeds, E. (1976) 'Accounting for behavioural differences: three political systems and the responses of squatters in Brazil, Peru and Chile', in J. Walton and L.H. Masotti (eds) *The City in Comparative Perspective*, New York: Halsted Press, pp. 193–248.

de Leeuw, F. and Ozanne, L. (1981) 'Housing', in H. Aaron and J. Pechman (eds) *How Taxes Affect Economic Behavior*, Washington, DC: The Brookings Institute.

Lewis, O. (1959) *Five Mexican Families*, New York: Basic Books.

Lewis, O. (1968) *La Vida: A Puerto Rican Family in the Culture of Poverty*, New York: Vintage Books.

Lim, G.C. (1988) 'Theory and taxonomy of sectoral, distributional, and spatial policies', *Environment and Planning, C, Government and Policy* 6, 2, 225–36.

Little, I.M.D. and Mirrlees, J.A. (1974) *Project Appraisal and Planning for Developing Countries*, London: Heinemann Educational.

Londero, E. (1987) *Benefits and Beneficiaries: An Introduction to Estimating Distributional Effects in Cost-Benefit Analysis*, Washington, DC: Inter-American Development Bank.

Loomis, J.B. (1989) 'Test-retest reliability of the contingent valuation method: a comparison of general population and visitor response', *American Journal of Agricultural Economics* 71, 1 (February), 76–84.

Loomis, J.B. (1990) 'Comparative reliability of the dichotomous choice and open-ended contingent valuation techniques', *Journal of Environmental Economics and Management* 18, 1 (January), 78–85.

Lundgren, T., Schlyter, A. and Schlyter, T. (1969) *Kapwepwe Compound – A Study of an Unauthorized Settlement in Lusaka, Zambia*, Lund: University of Lund.

McConnell, K.E. (1990) 'Models for referendum data: the structure of discrete choice models for contingent valuation', *Journal of Environmental Economics and Management* 18, 1 (January), 19–34.

McGoogan, E. (1984) 'The autopsy and clinical diagnosis', *Journal of the Royal College of Physicians of London* 18, 240–3.

McKenzie, R.B. and Tullock, G. (1981) *The New World of Economics: Explorations into the Human Experience*, Homewood, Ill.: Irwin.

McLeod, R. (1987) 'Low income shelter strategies in Kingston: solutions of the informal sector', Unpublished report, Regional Housing and Urban Development Office (RHUDO), USAID, Kingston.

Maddala, G.S. (1983) *Limited-Dependent and Qualitative Variables in Econometrics*, Cambridge: Cambridge University Press.

Maddala, G.S. (1988) *Introduction to Econometrics*, New York: Macmillan.

Maddala, G.S. and Trost, R.P. (1982) 'On measuring discrimination in loan markets', *Housing Finance Review*, 245–68.

Maia Gomes, G. (1986) *The Roots of State Intervention in the Brazilian Economy*, New York: Greenwood Press.

Malpezzi, S. (1984a) 'Rent controls: an international comparison', paper presented at the American Real Estate and Urban Economics Association Meetings.

Malpezzi, S. (1984b) 'Analyzing an urban housing survey: economic models and statistical techniques', UDD Discussion Paper No. 52, Washington, DC: The World Bank.

Malpezzi, S. (1985) 'Rent control and housing market equilibrium: theory and evidence from Cairo, Egypt', Unpublished PhD Dissertion, The George Washington University, Washington, DC.

Malpezzi, S. (1989) 'Analyzing incentives in housing programs: evaluating costs and benefits with a present value model', INU Discussion Paper No. 23, Washington, DC: The World Bank.

Malpezzi, S., Bamberger, M. and Mayo, S.K. (1981) 'Planning an urban housing survey: key issues for researchers and program managers in developing countries', UDD Discussion Paper No. 44, Washington, DC: The World Bank.

Malpezzi, S. and Mayo, S. (1985) 'Housing demand in developing countries', World Bank Staff Working Papers No. 733, Washington, DC: The World Bank.

Malpezzi, S. and Mayo, S.K. (1987) 'The demand for housing in developing countries: empirical estimates from household data', *Economic Development and Cultural Change* 34, 4 (July), 687–721.

Malpezzi, S., Mayo, S.K., Silveira, R. and Quintos, C. (1988). *Measuring the Costs and Benefits of Rent Control: Case Study Design*, Report INU 24, Policy Planning and Research Staff Discussion Paper, Infrastructure and Urban Development Department, Washington, DC: The World Bank.

Malpezzi, S., Tipple, A.G. and Willis, K.G. (1990) 'Costs and benefits of rent control: a case study in Kumasi, Ghana', World Bank Discussion Papers No. 74, Washington, DC: The World Bank.

Maniscalco, C.I., Doherty, M.E. and Ullman, D.G. (1980) 'Assessing discrimination: an application of social judgement technology', *Journal of Applied Psychology* 65, 284–8.

Marsh, R.M. (1967) *Comparative Sociology: a Codification of Cross-societal Analysis*, New York: Harcourt, Brace and World.

Martin, R.J. (1974) 'The architecture of underdevelopment or the route to self-determination in design', *Architectural Design* 10, 626–34.

Martin, R.J. (1983) 'Upgrading', in R.J. Skinner and M.J. Rodell (eds) *People, Poverty and Shelter: Problems of Self-Help Housing in the Third World*, London: Methuen.

Masotti, L.H. and Walton, J. (1976) 'Comparative urban research: the logic of comparisons and the nature of urbanism', in J. Walton and L.H. Masotti (eds) *The City in Comparative Perspective*, New York: Halsted Press, pp. 1–15.

Mayo, S. K., Malpezzi, S. and Gross, D.J. (1986) 'Shelter strategies for the urban poor in developing countries', *The World Bank Research Observer* 1, 183–203.

Meehl, P.E. (1954) *Clinical Versus Statistical Prediction*, Minneapolis: University of Minnesota Press.

Meehl, P.E. (1986) 'Causes and effects of my disturbing little book', *Journal of Personality Assessment* 50, 370–5.

Merrit, R.L. and Rokkan, S. (eds) (1966) *Comparing Nations: the Use of Quantitative Data in Cross-national Research*, New Haven: Yale University Press.

Ministry of Energy and Water Resources Development (Zimbabwe) (1985) *Water Tariff Study*, National Master Plan for Rural Water Supply and Sanitation, Harare, Zimbabwe.

Ministry of Planning, Shelter Unit, H.K. of Jordan (1987) *The 1987 National Housing Survey*, Amman: Ministry of Planning, Shelter Unit.

Mishan, E.J. (1982) *Cost Benefit Analysis*, London: George Allen and Unwin.

Mitchell, R. and Carson, R.T. (1989) *Using Surveys to Value Public Goods: The Contingent Valuation Method*, Washington, DC: Resources for the Future.

Morgenstern, O. (1963) *On the Accuracy of Economic Observations*, Princeton: Princeton University Press.

Mubanga, A.M. (1979) 'Chaisa and George essential resettlement: "The Drop-outs"', Lusaka: Lusaka Housing Project Evaluation Team Working Paper 26.

Mulenga, A.D. (1978) 'A descriptive analogy of the physical resettlement process', Lusaka: Lusaka Housing Project Evaluation Team Working Paper 24.

Muller, M.S. (1979) *Chawama – To Make a Good Place Better: The Socio-Economic History of a Squatter Settlement in Lusaka, Zambia*, London: Development Planning Unit, University College.

Murray, M. (1975) 'The distribution of tenant benefits in public housing', *Econometrica* 43, 4, 771–88.

Nicholson, R.J. and Topham, N. (1971) 'The determinants of investment in housing by local authorities: an econometric approach', *Journal of the Royal Statistical Society Series A*, 134, 273–303.

NHT (no date) (National Housing Trust of Jamaica) 'Objectives of the NHT', Internal document, NHT, Kingston.

NHT (1988a) Mortgagor Master File (computer tape, data as of May 1988), Kingston: NHT.

NHT (1988b) Contributor File, 1981 (computer tape), Kingston: NHT.

Nientied, P. and Van Der Linden, J. (1985) 'Approaches to low-income housing in the third world: some comments, *International Journal of Urban and Regional Research* 9, 3 (September).

Nowak, S. (1989) 'Comparative studies and social theory', in M.L. Kohn (ed.) *Cross-national Research in Sociology*, New York: Sage Publications: 34–56.

Oakley, D. and Raman, U.K. (1965) *The Rural Habitat – Dimensions of Change in Village Homes and House Groupings*, New Delhi: School of Planning and Architecture.

Offe, C. (1975) 'The theory of the capitalist state and the problem of policy formation', in L. Lindberg *et al.* (eds) *Stress and Contradiction in Modern Capitalism*, Lexington: Heath.

Oliver, P. (ed.) (1969) *Shelter and Society*, London: Barrie and Rockliffe.

Oliver, P. (ed.) (1971) *Shelter in Africa*, London: Barrie and Jenkins.

Oliver, P. (ed.) (1975) *Shelter, Sign and Symbol*, London: Barrie and Jenkins.

Oliver, P. (1987) *Dwellings: The House across the World*, Oxford: Phaidon.

Olsen, E. (1972) 'An econometric analysis of rent control', *Journal of Political Economy* 80 (November), 1081–100.

Ozanne, L. and Malpezzi, S. (1985) 'The efficacy of hedonic estimation', *Journal of Economics and Social Measurement* 13, 2, 153–72.

Papps, I. (1987) 'Techniques of project appraisal', in N. Gemmell (ed.) *Surveys in Development Economics*, Oxford: Blackwell, pp. 307–40.

Pastore, J. and Skidmore, T.E. (1985) 'Brazilian labor relations: A new era?' in H. Juris, M. Thompson and W. Daniels (eds) *Industrial Relations in a Decade of Economic Change*, Madison, Wis.: Industrial Relations Research Association, pp. 73–113.

Pearce, D.W. and Nash, C.A. (1981) *The Social Appraisal of Projects*, London: Macmillan.

Peattie, L.R. (1982) 'Some second thoughts on sites and services', *Habitat International*, 6, 1/2, 133–9.

Peil, M. (1976) 'African squatter settlements: a comparative study', *Urban Studies* 13, 2, 155–66.

Peil, M. and Sada, P.O. (1984) *African Urban Society*. Chichester: John Wiley.

Phillips, C.V. and Zeckhauser, R.J. (1989) 'Contingent valuation of damage to natural resources: how accurate? how appropriate?' *Toxics Law Reporter* 4 (October), 520–9.

Pickvance, C. (1981) 'Policies as chameleons: an interpretation of regional policy and office policy in Britain', in M. Dear and A. Scott (eds) *Urbanization and Urban Planning in Capitalist Society*, New York: Methuen, pp. 231–65.

Portes, A. (1984) 'Latin American urbanisation in the years of the crisis', *Latin American Research Review*, 24, 3, 7–44.

Ragin, C. (1987) *The Comparative Method: Moving Beyond Qualitative and Quantitative Strategies*, Berkeley, Calif.: University of California Press.

Ragin, C. (1989) 'New directions in comparative research', in M.L. Kohn (ed.) *Cross-national Research in Sociology*, New York: Sage Publications: 57–76.

Ragin, C. and Zaret, D. (1983) 'Theory and method in comparative research: two strategies', *Social Forces* 61, 731–54.

Rakodi, C. (1978) 'George 1976: initial results of the first preliminary sample survey – operation and policy implications', Supplement to Working Paper No. 4, Lusaka: Lusaka Housing Project Evaluation Team.

Randall, A., Grunewald, O., Johnson, S., Ausness, R. and Pagoulatos, A. (1978) 'Reclaiming coal surface mines in Central Appalachia: a case study of the benefits and costs', *Land Economics* 54, 4, 472–89.

Randall, A., Hoehn, J.P. and Brookshire, D. (1983) 'Contingent valuation surveys for evaluating environmental assets', *Natural Resources Journal* 23, 3, 635–48.

Randall, A., Ives, B. and Eastman, C. (1974) 'Bidding games for valuation of aesthetic environmental improvements', *Journal of Environmental Economics and Management* 1, 132–49.

Rapoport, A. (1967) 'Yagua – an Amazon dwelling', Landscape 16, 3 (Spring), 27–30.

Rapoport, A. (1969a) *House Form and Culture*, Englewood Cliffs, NJ: Prentice Hall.

Rapoport, A. (1969b) 'The Pueblo and the Hogan: a cross-cultural comparison of two responses to an environment', in P. Oliver (ed.) *Shelter and Society*, London: Barrie and Rockliffe, pp. 66–79.

Rapoport, A. (1975a) 'Australian Aborigines and the definition of place' in P. Oliver (ed.) *Shelter, Sign and Symbol*, London: Barrie and Jenkins: 38–51.

Rapoport, A. (1975b) 'Toward a redefinition of density', *Environment and Behaviour* 7, 2 (June), 133–58.

Rapoport, A. (1977) *Human Aspects of Urban Form*, Oxford: Pergamon.

Rapoport, A. (1978a) 'Culture and the subjective effects of stress', *Urban Ecology* 3, 3 (November), 241–61.

Rapoport, A (1978b) 'Nomadism as a man-environment system', *Environment and Behaviour* 10, 2 (June), 216–46.

Rapoport, A (1978c) 'Culture and environment', *Ecologist Quarterly* 4 (Winter), 269–79.

Rapoport, A. (1979) 'An approach to designing third world environments', *Third World Planning Review* 1, 1 (Spring), 23–40.

Rapoport, A (1980a) 'Cross-cultural aspects of environmental design', in I. Altman, A. Rapoport and J.R. Wohlwill (eds) *Environment and Culture*, Vol. 4 of *Human Behaviour and Environment*, New York: Plenum, pp. 7–46.

Rapoport, A. (1980b) 'Culture, site-layout and housing', *Architectural Association Quarterly* 12, 1, 4–7.

Rapoport, A. (1980c) 'Towards a cross-culturally valid definition of housing', in R.R. Stough and A. Wandersman (eds) *Optimising Environments (Research, Practice and Policy)*, (EDRA 11), Washington, DC: Environmental Design Research Association (EDRA).

Rapoport, A. (1980–81) 'Neighbourhood heterogeneity or homogeneity', *Architecture and Behaviour* 1, 1, 65–77.

Rapoport, A. (1982) 'Urban design and human systems: on ways of relating buildings to urban fabric', in P. Laconte, J. Gibson and A. Rapoport (eds) *Human and Energy Factors in Urban Planning – A Systems Approach*, The Hague, Nijhott: 161–84.

Rapoport, A. (1983a) 'Development, culture change and supportive design', *Habitat International* 7, 5/6, 249–68.

Rapoport, A. (1983b) 'The effects of environment on behaviour', in J.B. Calhoun (ed.) *Environment and Population: Problems of Adaptation*, New York: Praeger.

Rapoport, A. (1985) 'Thinking about home environments: a conceptual framework', in I. Altman and C.M. Werner (eds) *Home Environments*, Vol. 8 of *Human Behaviour and Environment*, New York: Plenum, pp. 255–86.

Rapoport, A. (1986a) 'The use and design of open spaces in urban neighborhoods', in D. Frick (ed.) *The Quality of Urban Life (Social, Psychological and Physical Conditions)*, Berlin: de Gruyter, pp. 159–75.

Rapoport, A. (1986b) 'Culture and built form – a reconsideration', in D.G. Saile (ed.), *Architecture in Cultural Change (Essays in Built Form and Culture Research)*, Lawrence: University of Kansas, pp. 157–75.

Rapoport, A. (1988) 'Spontaneous settlements as vernacular design', in C.V. Patton (ed.) *Spontaneous Shelter*, Philadelphia: Temple University Press, pp. 51–77.

Rapoport, A. (1989a) 'On the attributes of "tradition"', in J-P. Bourdier and

N. Al-Sayyad (eds) *Dwellings, Settlements and Tradition (Cross-Cultural Perspectives)*, Lanham, MD: University Press of America, pp. 77–105.

Rapoport, A. (1989b) 'On Regions and Regionalism', in N. C. Markovich, W.F. E. Preiser and F.G. Sturm (eds) *Pueblo Style and Regional Architecture*, New York: Van Nostrand Reinhold, pp. 272–88.

Rapoport, A. (1990a) *The Meaning of the Built Environment*, Tucson: University of Arizona Press (revised edn).

Rapoport, A. (1990b) *History and Precedent in Environmental Design*, New York: Plenum.

Rapoport, A. (1990c) 'Systems of activities and systems of settings' in S. Kent (ed.) *Domestic Architecture and the Use of Space*, Cambridge: Cambridge University Press.

Rapoport, A. (1990d) 'Environmental quality and quality profiles', in N. Wilkinson (ed.) *Quality in the Built Environment, Conference Proceedings, July, 1989*, Newcastle upon Tyne: Open House International Association.

Rawlings, J.O. (1988) *Applied Regression Analysis: A Research Tool, Pacific Grove, CA: Wadsworth and Brooks*.

Ray, A. (1984) *Cost-Benefit Analysis: Issues and Methodologies*, Baltimore: Johns Hopkins University Press.

Reynolds, C. and Carpenter, R. (1975) 'Housing finance in Brazil', *Latin American Urban Research 5*.

Ricketts, M. (1981) 'Housing policy: towards a public choice perspective', *Journal of Public Policy* 1, 501–22.

Ricketts, M. (1982) 'A politico-financial model of local authority rents and rate fund contributions in the UK', *Public Choice* 39, 399–414.

Ricketts, M. (1983) 'Local Authority housing investment and finance: a test of the theory of regulation', *The Manchester School* 51, 45–62.

Robben, P. and van Stuijvenberg, P. (1986) 'India's urban housing crisis: why the World Bank's sites and services schemes are not reaching the poor in Madras', *Third World Planning Review* 8, 4.

Robinson, R. (1981) 'Housing tax expenditures, subsidies and the distribution of income', *The Manchester School* 49, 91–110.

van Rooij, T. (1978) 'Support housing for the rented sector', *Open House* 3, 2, 2–11.

Rushinek, A. and Rushinek, S.F. (1987) 'Using financial ratios to predict insolvency', *Journal of Business Research* 15, 1 (February).

Rutter, A.F. (1971) 'Ashanti Vernacular Architecture', in P. Oliver (ed.), *Shelter in Africa*, London: Barrie and Jenkins.

Sadalla, E. K., Snyder, P.Z. and Stea, D. (1977) 'House form and culture revisited', in P. Suedfeld and J.A. Russell (eds) *The Behavioural Basis of Design*, (EDRA 7), Book 2, Stroudsburg, PA: Dowden, Hutchins and Ross.

Sarin, M. (1982) *Urban Planning in the Third World: The Chandigarh Experience*, Oxford: Mansell.

SAS (1986) *SUGI Supplemental Library User's Guide*, Version 5 edn, Cary, NC: SAS Institute.

Saunders, R. and Warford, J. (1977) *Village Water Supply*, Baltimore: Johns Hopkins University Press.

Sawhill, I.V. (1990) 'Tapping the Government's hidden subsidies', *The Urban Institute Policy and Research Report* 20, 1, 28.

Sayer, A. (1984) *Method in Social Science*, Dover: Hutchinson.

Sayer, A. (1985) 'The difference that space makes', in D. Gregory and J. Urry (eds) *Social Relations and Spatial Structures*, London: Macmillan, pp. 49–66.

Schildkrout, E. (1978) *People of the Zongo: The Transformation of Ethnic Identities in Ghana*, Cambridge: Cambridge University Press.

Schlyter, A. (1981) 'Upgraded George Revisited,' in C. Rakodi and A. Schlyter (eds), *Upgrading in Lusaka – Participation and Physical Changes*, Stockholm: Swedish Council for Building Research.

Schlyter, A. (1984) *Upgrading Reconsidered – The George Studies in Retrospect*, Gavle: The National Swedish Institute for Building Research.

Schlyter, A. (1987) 'Commercialisation of housing in upgraded squatter areas: the case of George, Lusaka, Zambia', *Trialog* (Darmstadt) 13/14, 24–29, and *African Urban Quarterly* 2, 3 (August, 1987), 287–97.

Schlyter, A. (1988) *Women Householders and Housing Strategies – The Case of George, Zambia*, Gavle: The National Swedish Institute for Building Research.

Schlyter, A. (1990) *Twenty Years in George: A Longitudinal Study of a Housing Area in Lusaka, Zambia*, Lund: The National Swedish Institute for Building Research.

Schlyter, A. and Schlyter, T. (1980) *George – The Development of A Squatter Settlement in Lusaka*, Zambia, Stockholm: Swedish Council for Building Research.

Schon, D.A. (1983) *The Reflective Practitioner: How Professionals Think in Action*, New York: Basic Books.

Schteingart, M. (1988) 'Presentacion', *Medio Ambiente y Urbanizacion 24*, 1–2.

Schwerdtfeger, F.W. (1982) *Traditional Housing in African Cities: a Comparative Study of Houses in Zaria, Ibadan and Marrakech*, Chichester: John Wiley.

Seeling, M.Y. (1978) *The Architecture of Self-Help Communities*, New York: Architectural Record Books.

Siegal, S. (1956) *Nonparametric Statistics for Behavioral Sciences*, New York: McGraw-Hill.

Silitshena, R.M.K. (1982) 'Population movement and settlement patterns in contemporary Botswana', in R.R. Hitchcock and M.R. Smith (eds) *Settlement in Botswana*, Johannesburg: Heinemann Educational.

Sinden, J.A. (1974) 'A utility approach to the evaluation of recreational and aesthetic experiences', *American Journal of Agricultural Economics* 56, 1 (February), 61–72.

Singini, R.E. (1978) 'George 1978: Primary Surveys I and II: comparisons and their operational and policy implications', Lusaka: Lusaka Housing Project Evaluation Team Working Paper 21.

Sinha, A. (1990) 'A critique of state self-help housing in Northern India', *Open House International* 15, 1.

Siqueira, W. (1985) *BNH Nao!*, Rio de Janiero: Federaçao Nacional dos Administradores de Empresas, e o Sindicato dos Técnicos de Administraçao do Rio de Janeiro.

Smith, V.L. (1980) 'Experiments with a decentralised mechanism for public good decisions', *American Economic Review* 70: 584–99.

Spradley, J. (1979) *The Ethnographic Interview*, New York: Holt, Rinehart and Winston.

Squire, L. and van der Tak, H.G. (1975) *Economic Analysis of Projects*, Baltimore: Johns Hopkins University Press.

STATIN (Statistical Institute of Jamaica) and World Bank (1988) *Living Conditions Survey: Jamaica*, Preliminary Report, November, Kingston: STATIN.

Stern, N.H. (1987) 'Aspects of the general theory of tax reform', in D.M.G. Newbery and N.H. Stern (eds), *The Theory of Taxation for Developing Countries*, Oxford: Oxford University Press, pp. 60–91.

Storey, D., Keasey, K., Watson, R. and Wynarczyk, P. (1987) *The Performance of Small Firms*, London: Croom Helm.

Stren, R.E. (1978) *Housing the Urban Poor in Africa: Policy, Politics, and Bureaucracy in Mombasa*, Berkeley Calif.: University of California Institute of International Studies.

Struyk, R. (1988) 'Understanding high vacancy rates in a developing country: Jordan', *Journal of Developing Areas* 22, 3, 373–80.

Taffler, R.J. (1982) 'Forecasting company failure in the UK using discriminant analysis and financial ratio data', *Journal of the Royal Statistical Society, Series A* 145, 342–58.

Tamari, M. (1978) *Financial Ratios: Analysis and Prediction*, London: P. Elek.

Tewari, V.K. and Kumar T.K. (1986) *Rent Control in India: Its Economics Effects and Implementation in Bangalore*, Washington, DC: The World Bank, Water Supply and Urban Development Department Discussion Paper UDD–91.

Thibodeau, T. (1981) *Rent Regulation and the Market for Rental Housing Services*. Washington, DC: Urban Institute Research Paper 3090–1.

Thomson, K.J. and Willis, K.G. (1986) 'Errors-in-variables: a problem in regression and its solution', *Environment and Planning A* 18, 687–93.

Thrift, N. (1985) 'Research policy and review 1. Taking the rest of the world seriously? The state of British urban and regional research in a time of economic crisis', *Environment and Planning A* 17, 7–24.

Tipple, A.G. (1983) *Asante Culture and Low Income Housing Policies: An Examination of Antithesis*, Urban Development Department Discussion Paper UDD–45, Washington, DC: The World Bank.

Tipple, A.G. (1986) 'A revolution in property rights (Ghana)', *West Africa*, 27 January, 179–80.

Tipple, A.G. (1987) *The Development of Housing Policy in Kumasi, Ghana, 1901 to 1981*, Newcastle upon Tyne: Centre for Architectural Research and Development Overseas, University of Newcastle upon Tyne.

Tipple, A.G. (1988) *The History and Practice of Rent Control in Kumasi, Ghana*, Urban Development Division Working Paper No. 88–1, Washington, DC: The World Bank.

Turner, J.F.C. (1968) 'Housing priorities, settlement patterns, and urban development in modernising countries', *Journal of the American Institute of Planners* 34, 354–63.

Turner, J.F.C. (1976) *Housing by People: Towards Autonomy in Built Environments*, London: Marion Boyars.

Turner, M.A. (1990) *Housing Market Impacts of Rent Control: The Washington, DC, Experience*, Washington, DC: Urban Institute Report 90–1.

Tversky, A. and Kahneman, D. (1974) 'Judgement under uncertainty: heuristics and biases', *Science* 185, 1124–31.

Tversky, A., Sattah, S. and Slovic, P. (1988) 'Contingent weighting in judgement and choice', *Psychological Reviews* 95, 3, 371–84.

United Nations (1977) *Formulation of National Settlement Policies and Strategies* Nairobi: UNCHS (Habitat).

United Nations (1980) *UN Seminar of Experts on Building Codes and Regulations in Developing Countries*, Nairobi and Stockholm: UNCHS (Habitat) and Swedish Council for Building Research.

United Nations (1982) *Development of the Indigenous Construction Sector*, Nairobi: UNCHS (Habitat).

United Nations (1984) *Upgrading of Inner-City Slums*, Nairobi: UNCHS (Habitat).

United Nations (1987a) *Global Report on Human Settlements*, London: Oxford University Press for UNCHS (Habitat).

United Nations (1987b) *Standards and Specifications for Local Building Materials*, London: Intermediate Technology Publications on behalf of UNCHS (Habitat).

United Nations (1989a) *Expert Group Meeting on Rental Housing in Developing Countries: a Review of the Present Situation*, Unpublished Report: UNCHS (Habitat).

United Nations (1989b) *Strategies for Low-Income Shelter and Services Development: the Rental-Housing Option*, Nairobi: UNCHS (Habitat).

Upton, G.J.G. (1989) 'The analysis of cross tabulated survey data', in S. Conrad (ed.) *Assignments in Applied Statistics*, Chichester: Wiley.

Urban Affairs Quarterly (1975) 'The City in Comparative Perspective', (September).

Vaughn, W.J. and Russell, C.S. (1982) 'Valuing a fishing day: an application of a systematic varying parameter model', *Land Economics* 58, 4 (November), 450–63.

Vayda, A.P. (1983) 'Progressive contextualisation: methods for research in human ecology', *Human Ecology* 37, 2, 325–43.

Walton, J. (1977) *Elites and Economic Development: Comparative Studies of the Political Economy of Latin American Cities*, Austin: University of Texas Press.

Walton, J. and Masotti, L.H. (eds) (1976) *The City in Comparative Perspective*, New York: Halsted Press.

Ward, P.M. (ed.) (1982) *Self-Help Housing: a Critique*, London: Mansell.

Ward, P. (1986) *Welfare Politics in Mexico: Papering Over the Cracks*, The London Research Series in Geography 9, Boston: Allen and Unwin.

Watson, D.S. and Holman, M.A. (1977) *Price Theory and its Uses*, Boston: Houghton-Mifflin.

Webber, R. and Craig, J. (1978) *Socio-economic Classification of Local Authority Areas*, Studies on Medical and Population Subjects No. 35, Office of Population Censuses and Surveys, London: HMSO.

Webber, R.J. (1978) 'Making the most of the census for strategic analysis', *Town Planning Review* 49, 274–84.

Weston, J.F. and Brigham, E.F. (1972) *Managerial Finance*, London: Dryden Press.

Westwick, C.A. (1987) *How to use Management Ratios*, Aldershot: Gower.

Whittington, D., Lauria, D.T. and Mu, X. (1990) 'Paying for urban services: a study of water vending and willingness to pay for water in Onitsha, Nigeria', PRE Working Paper No. 363, Infrastructure and Urban Development Department, Washington, DC: The World Bank.

Whittington, D., Smith, V. K., Okarafor, A., Okore, A., Liu, J. L., Ruiz, L.K. and McPhail, A. (1990) 'Giving respondents time to think in contingent valuation studies', PRE Working Paper, Infrastructure and Urban Development Department, Washington, DC: The World Bank.

Whittington, D., Mujwahuzi, M., McMahon, G. and Choe, K. (no date) *Willingness to Pay for Water in Newala District, Tanzania: Strategies for Cost Recovery*, USAID Water and Sanitation for Health Project, Report on Activity No. 445, Washington, DC: USAID.

Whyte, W.F. (1984) *Learning from the Field: A Guide from Experience*, New Delhi: Sage.

Wikan, U. (1980) *Life among the Poor in Cairo*, London: Tavistock.

Willis, K.G. (1972) 'Rural development planning and policy evaluation', *Geografiska Annaler Series B*, 54, 109–16.

Willis, K.G. (1974) *Problems in Migration Analysis*, Farnborough: Saxon House.

Willis, K.G. (1980) *The Economics of Town and Country Planning*, London: Collins (Granada).

Willis, K.G. (1987) 'Discriminant analysis', in C.S. Yadav (ed.) *Models in Urban Geography, Part II (Mathematical)*, Perspectives in Urban Geography Volume 4-B, New Delhi: Concept Publishing.

Willis, K.G. and Benson, J.F. (1988) 'A comparison of user benefits and costs of nature conservation at three nature reserves', *Regional Studies* 22, 417–28.

Willis, K. G., Malpezzi, S. and Tipple, A.G. (1990) 'An econometric and cultural analysis of rent control in Kumasi, Ghana', *Urban Studies*, 27, 2: 241–58.

Wonnacott, R.J. and Wonnacott, T.H. (1970) *Econometrics*, New York: Wiley.

Wonnacott, R.J. and Wonnacott, T.H. (1985) *Introductory Statistics*, 4th edn, New York: Wiley.

Woodfield, A. (1989) *Housing and Economic Adjustment*, New York: Taylor and Francis, for the Department of International Economic and Social Affairs, United Nations.

World Bank, (1985) *Kenya: Economic Development and Urbanization Policy*, Report No. 4148-KE, Washington, DC: The World Bank.

Yellen, J. (1985) 'Bushmen', *Science 85* (May): 40–8.

Yotopoulos, P. and Nugent, J. (1976) *Economics of Development: Empirical Investigations*, London: Harper and Row.

Zambia, Government of (1973) *Mwaziona: A Study of an Unofficial Housing Area*, Lusaka: Department of Town and Country Planning.

Zambia, Central Statistical Office (1985) *1980 Census of Population and Housing*, (preliminary report and specially ordered data sheets of enumeration areas in George).

Index

analysis 210–11; shelter
investment decisions 211
discriminant analysis 13, 126–42;
determinants of tenure choice
129–31; housing service choice in
Kumasi 127–9; other applications
of 139–41; predicting changes in
tenure choice 137–9
Dixon, J.A. (in Hufschmidt, M. *et al.*)
247
Doherty, M.E. (in Maniscalto, C.I.
et al.) 167
Doling, J.F. 140
de Dombal, F.T. 5, 167; Leaper, D.J.,
Horrocks, J.C., Staniland, J.R. and
McCann, A.P. 9, 167
Draper, N.R. and Smith, H. 145
Dreze, J. and Stern, N.A. 235
DSI *see* debt-service over income
ratio
Dunn, G. (in Everitt, B.S. *et al.*) 126
Durand, D. 140

'ecological fallacy', the 104
econometric analysis 10, 169–87;
conceptual framework 171–2;
maintenance 186–7; measuring
the inputs 172–5; profitability
180–3; results 175–80; trends in
the number of units 183–6
economically weaker section (EWS)
20, 21, 23, 29
'edifice complex', the 108
Edwards, M. 1
emic and etic categories 32, 34n
environment-behaviour relations
(EBR) *see* cultural change analysis
ethical problems 84–5
ethnicity 49
evaluation, housing 17–19
Evans, P.B., Rueschemeyer, D. and
Skocpol, T. 97
Everitt, B.S. and Dunn, G. 126
extended interviews *see* participant
observation; time series analysis
external benefits 245–7

'family housers' 128–9
Fathy, H. 8
Feber, R. and Hirsch, W.Z. 8

FGAL *see* financing gap left after
housing loan
financing gap left after housing loan
(FGAL) 114, 116–24
Fleming, M.C. and Nellis, J.G. 10
Fogerty, M.S. 140
Follain, J. and Jiminez, E. 247
Franck, K. 26
Freeman, A.M. 190
Friedman, J., Jimenez, E. and Mayo,
S.K. 157, 166
front-back relationships 39, 56–7

garden cities 8
Geiger, P.P. and Davidovich, F.R.
105, 106
George study, a review of the 62–79;
methods of investigation 65–8;
occupancy rates and
modernisation 75–8; urban
pattern and density 68–74
Ghana 15; controlled rental housing
223–30; determinants of
overcrowding 143–68; tenure
choice in Kumasi 126–42
Gilbert, Alan G. xiii, 81–95, 130, 132,
138, 141; and Goodman, D. 106;
and Healey, Patsy xiii; and Varley,
Ann xiii, 86; and Ward, Peter M.
xiii, 86
Glaser, D. 5, 167
Gleaner (Kingston) 99, 100
Global Strategy for Shelter to the
Year 2000 (UN) 2
Goetz, M.L. and Wofford, L.E. 260
Golding, B. 99–100
Goodman, D. (in Gilbert, Alan G.
et al.) 106
government mortgage records,
analysis of 12–13, 96–112;
accessing data 101–3; advantages
of using 103–6; disadvantages of
using 106–9; mortgage subsidies
109–11; state theory and the
housing crisis 97–101; suggestions
for research directions 111–12
Graham, D. and Pollard, S. 106
Graves, P., Murdoch, J.C., Thayer,
M.A. and Waldman, D. 10
Grose, R.N. 99–100